SOFT DECORATION
FABRICS IN HOME DESIGN

宅妆——家居软装饰

杜丙旭 编　　代伟楠　李婵 译

辽宁科学技术出版社

Contents
目录

004 **Soft Space, Taste of Life**
软化空间，品味生活

007 **Soft Decoration: A Lifestyle**
软装营造生活

008 American Style 美式风情

048 European Style 欧式风格

092 Chinese Classical Style 中式经典

104 Mix and Match 混搭

147 **Soft Decoration in Home Space**
家居软装饰

148 Living Room 起居室

162 Dining Room 餐厅

172 Bedroom 卧室

180 Washroom 洗手间

186 Entertainment and Leisure 娱乐与休闲

193 **Soft Decoration Details**
软装细节

194 Furnishings 陈设

208 Fabrics 织物

216 Texture 肌理

222 Project 1:
Notting Hill, London

案例1：伦敦诺丁山别墅

228 Project 2: House

案例2：房子

236 Project 3:
The Manor, 10 Davies Street

案例3：戴维斯大街10号庄园公寓

246 Project 4: Villa No.10

案例4：10号别墅

254 **Index** 索引

Soft Space, Taste of Life

软化空间，品味生活

赵牧桓
牧桓设计+灯光设计研究室 设计总监

Nowadays people begin to pay more and more attention to residential interior design, which clearly indicates the increasing importance of our home living quality.

Good circulation and aesthetic "interior hardware" are indispensable for comfortable home spaces. A good circulation design will help create a warm ambience, and a piece of artistic "hardware" will welcome you home everyday with a delightful mood.

However, in home interior design there's something that is often ignored, or even unprofessionally treated. That is decoration, or the popularly so-called "soft decoration". Soft decoration usually involves visual, tactile and sometimes olfactory senses.

A good soft decoration will not only complement the "hardware" of an interior in terms of diversity and comfort, but also reveal the taste and personality of the master. Unfortunately, more often than not, home decoration is treated in an unprofessional way. It is harmful, instead of helpful, to the aesthetic "hardware" an interior designer has created, and even will visually disturb the entire space. A situation a designer often encounters is that he spares no effort in creating an enjoyable living space, and gives it to the client who relentlessly destroys it with farfetched soft decoration. Hence we say the re-establishment of clients' taste is a rather practical problem. The client should have due respect to the expertise of designers, and believe in and take their advice, in order to have a harmonious environment at home. However, that's not something a designer could completely control – after all, he won't be involved in the client's life after completing the "hardware" of the house. In this sense, I think the re-creation of aesthetics with soft decoration is the real challenge in home interior design.

This book is about soft decoration in home spaces. How to decorate a home is really something that deserves learning and requires some aesthetic accomplishment as well. I'm glad to see the book with such a collection of beautiful home interior projects, because for both professional designers and laymen who are willing to improve their living quality, it would be an inspiring and valuable reference. All in all, I hope everyone could find his paradise when returning home everyday.

Hank M. Chao in Shanghai

住宅居家设计越来越受到国人重视。显示重视自己居家环境品质的人日益增多。

良好的动线设计和硬件上的美感可以说是住家设计的一种刚性需求。聪明的动线规划可以促进家里的和谐气氛，当然，一个具有美感的硬件让人每天回家会感到赏心悦目。

住宅设计里另一个比较容易忽视的环节，甚至比较会被不专业的处理的部分，就是装饰部分，也就是大多数人所谓的软装。软装涉及视觉、触觉和部分嗅觉的感官刺激。

好的软装势必会帮硬装添加更多的丰富性和舒适度，也可以增显主人的品位和气质。但很可惜这部分往往会被很不专业的处理，让原本很好的硬件不能好好的得到彰显，甚至让整体的视觉环境更加紊乱。这也是大多数设计师面临的问题：往往设计师把硬件部分交托给业主之后，然后业主亲自再把空间给摧毁了。于是，就产生了一个具体的问题：业主自身品位的再教育是很重要的，这部分业主也应该适度尊重设计师的专业性，听取并仰赖设计师的建议，这样比较能够让整体更加完整。这部分因为设计师的主控权相对薄弱，毕竟，设计师在交付业主硬件后不再参与业主日后的生活，所以，我认为"二度美感的再造"确实才是住宅设计真正的挑战的开始。

"宅妆"这个名字取得很好。如何点妆一个住家并帮她增添艳丽确实是一个大学问，也需要一定的美学素养。辽宁科学技术出版社愿意在这部分投入心力整理一些好的案例供专业与非专业但有兴趣让自己家里的居住品质有所提升的人士作为参考观摩，这点是值得肯定嘉许的。这点是身为设计者的我无法做到的。抛砖引玉，希望每个人都能够回家的时候，找到属于自己的那一片天。

赵牧桓于上海

Soft Decoration: A Lifestyle
软装营造生活

The ever-changing life brings differences to men, men's aesthetic appreciation and then the style of home design. In this sense, there is no everlasting design. Design is life and the designing process is a life experience; therefore design should undergo continuous improvement and development as life changes. In soft home design, it is unavailing to strive to make a project a perpetual classic work. Only by regarding soft decoration as a part of the soft space and the taste of life with fewer utilitarian purposes can one truly approach design and life themselves.

Life is not only an experience but a process of the accumulation of beauty. To understand the essence of life, one must know clearly what you want. Everyone has his or her own life, and thus the comprehension, dream and pursuit of beauty differ individually; a phenomenon also occurs in interior design: the requirement and aspiration for home decoration design vary from each other. In home decorative design, one needs to pay attention to so many aspects to achieve the amalgamation of decoration and life: the kind of ambience people want, the ways in which the design can win people's satisfaction and whether the designer is confident enough to put his decorative idea into practice.

Nowadays, soft home decoration, as an independent design category, has attracted more and more attention. The number of design corporations and designers engaged in soft decoration is increasing continuously. However, a large amount of superficial design can still be seen in most design cases, which look so farfetched and affected. Such kind of design simply focuses on the external changes of the interior space without any attention to the virtual and spiritual ones. It is just the outer form of soft decoration lacking the spiritual exploration.

We believe that fine soft decoration design should at least be endowed with such qualities as to meet the functional requirements of space, express beauty and its understanding of beauty in the aesthetic ambience and serve as a shocking power emitting art and elegance. Nevertheless, these are not enough, as the excellent design can both highlight the identity of the client and express his understanding of life. The precious metal, advanced furniture and rare material may not be applied in all soft home decoration but they are the necessary components of luxury space. The problem about how these can be achieved in effective means must be solved in soft home design. Good design must also walk into the inner world of human beings by carefully studying their eagerness for space. It is the home decoration which combines the unique definition of home design, men's feelings, demands, aspirations and dreams, variations and endurance into the space, regarding it as an element of human beings, can roll space and men into one and then fill the habitants' inner hearts with a sense of belonging and happiness. However, it is difficult to amalgamate men and the space just as there is always a single-step distance between aspirations and realities. The real difficulty of design lies in how to put ideas into practice and how to accomplish it besides the conception and inspiration of design.

生活会发生变化，人会变，人的审美取向会变，对应的家居设计风格也会发生变化，永恒的设计是不存在的。设计即是生活，设计的过程就是在体验生活，设计要跟随生活的变化不断进步和发展。对于家居软装饰，如果力图使一个项目的设计成为永恒的经典之作，很多努力就都会是徒劳的。应当将软装饰设计作为软化空间、品味生活的一部分，只有这样才是接近了设计本身，接近了生活本身。

生活既是体验，同时它又是一种积累——对于"美"的积累。要读懂生活，首先要明确的是自己想要的是什么。每个人都有自己的独特生活、对于"美"的向往和追求。反映到室内设计领域，对于家居装饰设计的要求和愿望也是不同的。在家居装饰设计中，让装饰与生活融合，需要更多关注人们想要看到周围的环境是什么，软装设计如何能够轻易地使人获得满足感，作为软装设计师是否有足够的信心将自己的理想付诸实践。

如今，家居软装饰设计越来越被人们重视并成为独立的设计门类。这一领域中也涌现出越来越多的专业软装设计师和设计公司。然而，在大多数的设计案例中，我们依然会看到那些停留在表象的设计，无论空间被设计成的风格如何，给人的感觉总是有些牵强、做作，似乎是刻意的流露。这样的设计被单纯地停留在室内空间外观的变化，而非内在的、实质的改变。只具备了软装的躯壳，而缺失软装的灵魂。

我们认为，好的软装设计除了要满足空间对于功能的需要，展现美观，空间洋溢着美学氛围和对于美的理解及其表现力，同时还要给人以艺术、高雅的震慑力，如果仅仅满足这些仍然是停留在表面的设计。好的设计必须要能够走进人们的心灵，需要更多的去研究人们对于空间的渴望，对于"家居"的独特定义。他的感受，他的需求，他的渴望与幻想，他的变化与恒久将人融化于空间，将空间作为人的一个因素，这样的家居装饰才是将空间与人合二为一，才能够让人升华出归宿感和温馨的幸福感。

空间与人合一的实现总是会存在一定的难度，就好像现实与美好之间总是会存在着一步之遥。设计中的一步之遥更多的除了取决于设计的构思，更重要的是要解决构想的现实性和最大实现的可能性。好的设计既要彰显出业主的身份，又要表现出其对生活的理解。比如说并非所有的家居软装中都要涉及贵重金属、高档家具、稀有材料等这些元素的应用，但要打造奢华的空间气质，这些却又必不可少，如何通过有效的手段和途径去实现这些愿望，是家居装饰设计中需要解决的。

American Style
美式风情

Featuring mostly the soft decoration and the auxiliary hard decoration, the American style decoration derives from the country lifestyle of western America. The characteristic of American style lies in the randomness in the classical style which means that it abandons the over-elaborate excessive decoration and extravagance while still possessing the elegant style of Classicalism and the functional allocation of Neoclassicism.

美式风情是以软装饰为主，硬装饰为辅的装饰风格，它是从美国西部乡村生活方式的装饰风格演变而来。美式风情的特点是在古典中略带随意，摒弃过度修饰的繁琐和奢华，却兼具古典主义的优美造型与新古典主义的功能配备。

The light blue coffee table seems old with traces of time. Echoing it, the grey sofa and brown cushions make the room calm and comfortable.

淡蓝色的方形茶桌透露出岁月的痕迹,与其相呼应的灰色沙发以及咖啡色的靠垫将空间烘托得沉稳而又安逸。

The brownish red wooden table structure, the black leather sofa, the cushions, carpet and furnishings such as the gallipot and plant, complete the American-style living room.

棕红色的木质框架茶几搭配黑色的皮质沙发，再点缀以靠垫、地毯等织物以及陶罐、绿植等陈设品，空间的美式风格被渲染得十分浓郁。

In American-style decoration, cloth sofas are always adopted accompanied with a lot of cushions to highlight the fabric texture, offering a warm and soft visual experience. Meanwhile, wooden coffee tables are often used as the main furniture in the space.

美式风格的装饰采用传统的布艺沙发，运用大量的靠垫增加装饰中的织物分量，给人以温暖柔和的视觉感受。于此同时，空间中的茶几等木质陈设物品成为装饰中的"骨架"，增加空间的装饰力度。

The commodious French window makes the space broader and brighter. To match it, sofas and lamps are chosen as the furnishings in the design of the living room. The power of soft decoration is manifested by the characteristics of the furniture itself rather than the miscellaneous decoration. The exquisite decorative treatment can be demonstrated through the design of the furnishings details such as the carved decorative patterns on the sofa, the gauze table cloth and the pendants of the lampshade.

宽敞的落地窗使空间显得开阔而明亮，与之相搭配，起居室的设计采用沙发和台灯作为陈列的内容。抛弃了繁琐的装饰，用家具自身的特点表现空间的软装力量。沙发的雕刻花纹、纱质的桌布、灯罩垂帘的设计等都透露出细节上的装饰处理，小处见大。

As the typical American furniture, the brownish leather sofa is big and wide, thick and solid, leaving people a warm feeling. The assorted wood tea table can also demonstrate some classical elements. The old metal ornamental lamppost evokes people's endless imagination.

棕黑色的皮质沙发宽大、厚实，典型的美式家具能够给人温暖的感觉。与其搭配的木质茶几能够透露出些许经典的元素。金属的、古老的灯台饰品令人产生无限的遐想。

The brown solid wood TV stand is also used as the display stand of articles, which can not only be used for the collection of articles but produce a commendable decorative effect with furnishings displayed on it. The old copper hinge, as a classical element, radiates intensely the nostalgia feelings while the LCD television with broad screen leads us back to the modern space. The soft decoration perfectly integrates Classicism with Modernism.

棕色的实木电视柜同时又作为物品的展示架,将陈设品陈列其上,既是收纳又能起到很好的装饰效果。铜质的老式合页作为经典元素散发出浓重的怀旧气息,而宽大屏幕的液晶电视则告诉我们空间是现代生活的空间。通过软装饰设计将经典与现代完美融合。

The spatial design focuses on the tea table surrounded by the American style sofas, behind which the wall is designed into a showcase and a TV stand that are made of wood in a bid to manifest the spatial steadiness and elegance.

此空间的设计以茶几为中心,将富于美式风格的沙发围绕四周,沙发的后面对应的墙壁设计成陈列柜和电视柜,均采用木质材料作为装饰元素,体现空间的厚重、雅致。

The soft design of the American style living room needs to be magnificent and steady to match up with the style of the commodious French window, which can be demonstrated not only by the unique appeal of American furniture but through the collaboration between each other. Sometimes, the simplified decoration, however, can make the space open, natural and gorgeous.

与室内宽大的落地窗相配合，美式起居室的软装设计同样要大气、沉稳。这一方面要靠美式家具独特的韵味来表现，而更多的则需要通过彼此间的搭配。有时候简化装饰反而能够使空间显得开阔、自然、大气。

In American style soft design, it is feasible to create a plane and casual decorative effect by abandoning the carved patterns on the furniture or the luxuriant decorative elements. The soft decoration in this project leads people to feel the American-style modest luxury by refining complexity into simplicity.

在美式风情的软装设计中，可以抛弃雕刻花纹或者采用过多的装饰元素，将空间打造成质朴的、不经意的装饰效果。本项目中的软装设计，将复杂提炼为简约，在雅致中让人体会美式低调的奢华。

The dark brown wooden coffee table surrounded by beige cloth sofas is well echoed by the two dark brown cabinets positioned on the two sides of the door. The space looks quiet and cosy.

深棕色的木质茶几被米色的布艺沙发所围合,房门两侧的深棕色陈列柜与茶几相呼应,空间显得宁静、悠闲。

In the soft decoration of some living rooms, the absence of brilliant decorative elements, however, makes the simple decorative style more respectful and closer to life whether in terms of a lamp or a wall hanging, which conveys the slight elegance in silence.

在一些起居室空间的软装饰中,没有耀眼的装饰元素,却在朴实中更加贴近生活,更加让人肃然。一盏灯,一个墙饰,在无声中传递出淡淡的优雅。

The pale blue cloth sofas go well with the carpet, creating a calming ambience. The small red leather sofa and the metal chandelier bring an elegant and noble air to the space.

淡蓝色的布艺沙发与地毯相搭配,渲染出空间淡然的情绪。红色的皮质沙发与屋顶的金属吊灯则又增添了空间些许的高贵气息。

In American style design, the decorative patterns may be chosen to manifest the rural feeling. In this project, the fabric sofas in collaboration with the plant patterns create a gentle and romantic spatial ambience. The patterns of the carpet work in concert with those of the wallpaper and the curtain and then jointly set off the intense decorative effect of the space with the help of the carvings on the supports of the tea table and the suspended ceiling.

在美式风情的设计中，可以采用花纹的图案来表现田园的感觉。在这个项目中采用布艺的沙发配合植物图案，营造出柔软、浪漫的空间气息。地毯的花纹与壁纸、窗帘的花纹相呼应，与茶几支架的雕刻、吊顶的图案雕刻共同渲染出空间的强烈的装饰效果。

The American idyllic living room uses the floral patterns as its decorative elements. The floral patterns, as lively decorative elements of the space, may leave people a messy visual sensation while forming an elegant dance rhythm.

田园风格的美式起居室，采用碎花的图案作为装饰元素。碎花图案容易给人以凌乱的视觉感受，同时也会形成舞动的曼妙之感，是空间中"活"的装饰元素。

The wooden sofas highlight the "hard" aspect of American style, while textiles such as the cushions, carpet and curtains, and green plants represent the "soft" aspect.

木质的沙发椅突出美式元素中刚硬的部分,而靠垫、地毯、窗帘等织物以及空间的绿植则凸显了美式风格柔美的一面。

This is an American-style dining room. The dining table, little chairs and stools are all made of wood, in a quietly elegant colour scheme.

充满美式风情的餐厅空间,餐桌、餐椅、餐凳均采用实木制作,色彩上突出淡雅、明亮。

The design of the dining room utilises natural timber as the component elements of the dining table and dining chairs to achieve the decorative effect which makes men closer to nature, recovering their original simplicity, and meanwhile creates the intense idyllic feelings.

餐厅的设计运用天然的木料作为餐桌、餐椅的组成元素起到装饰的效果,进而让人更加接近自然,给人以返璞归真的感觉,同时又营造出浓郁的田园风情,厚重中体现些许沉稳。

The round dining table, the leather dining chairs and the hanging lampshade of the pendant lamp are all decorated in the thematic colour—brown, which manifests dignity, steadiness and luxuriance. Both the metal candlesticks on the dining table and the paintings with modern elements on the wall become the finishing touch of the spatial soft decoration.

圆形餐桌、皮质就餐椅以及悬挂灯罩都采用了棕色为主题色，厚重、低沉、奢华。餐桌上的金属烛台以及墙壁上的现代元素的挂画，都成为空间软装的点睛之笔。

In the open dining space, the watercolour with stomachic dishes hanging at the hallway is showcased through the open door. The two dining chairs prepared in the dining room demonstrate extreme luxury.

开阔的就餐空间，开敞的房门将玄关处悬挂的大尺寸的、就餐内容的水彩画展现出来。餐厅只准备了两把就餐椅，贵气、奢华可见一斑。

An ivory-white small coffee table set beside a comfortable sofa, together with some green plants. Such is a typical American-style arrangement giving out a softly fragrant ambience.

沙发旁的小茶几,乳白色调透露出些许的柔和,摆上一束绿植,淡雅中增添温馨的氛围。

The wall in the space is designed into a tall bookshelf with books in various sizes on and two seats beside, which sets off the intense literary atmosphere.

空间的墙壁被设计成高大的书架,上面陈列着大大小小的书籍,两张座椅,将空间烘托出浓郁的书香气息,让人会自然地沉浸在书海之中。

Here is an American-style study. The leather sofa accompanied by the framed pictures makes the space feel dignified. The magnificent bookshelf further enhances the dignity.

美式风格的书房设计,皮质沙发与镶边的画框相辅相成,增加空间的厚重,高大的书架更凸显浓厚的书香气息。

The double door installed at the entrance, beside which there is a mirror in which the interior ornaments reflected enrich the decorative significance, is creatively designed in accordance with the original and ingenious design concept.

空间的入口处采用对开门的设计,在门的上部安装一面镜面,通过其反射出的室内的装饰物来为此入口增添装饰的意味,设计感强,而且理念独到、巧妙。

The commonly-seen simple vestibule design features a relatively small table with strong decorative sense on which there are several decorative items and a small mirror on the opposite wall, highlighting the typical style of the vestibule while possessing some practicalities.

比较常见的简单的玄关设计,采用一张不大但是装饰感浓烈的桌子,上面摆放一两件装饰物品,在对应的墙面再悬挂一面小镜,玄关的意味浓重而又具备一定的实用性。

The bed in dark brown, the bright-coloured bed sheet with floral patterns and the gauze curtains integrate gravity and liveliness into the same bedroom, which gives birth to the American pastoral feelings simultaneously.

深棕色的床，鲜艳的花床单以及纱质窗帘，将深沉与活泼统一在同一个卧室空间中，美式的田园风情油然而生。

The dark brown wooden bed and night table are designed with a vintage air, particularly the metal pull-tab. The carpet, curtains and beddings make the bedroom feel warm and soft.

深棕色的实木床体设计，配合实木床头柜、金属拉环显示出复古的气息，地毯、帷幔以及床上的织物增添了空间的柔软。

The soft decoration of this bed head comprises the design of the bedside cabinet, the lamp, the fresh flowers as well as the style design of the bed itself, using various colours such as brown, pink and green, which enrich the decoration of the bed head while increasing the spatial warmness.

这个床头的软装设计包括床柜、台灯、鲜花、挂画以及床头自身的造型设计,在色彩方面包括了棕色、粉色、绿色,使床头的装饰显得丰富,同时增添了空间的温馨之感。

Bed is not only the visual focus in the bedroom but the emphasis of the soft decoration of the bedroom. The soft decoration of the bed can set the basic tone of sentiments and the main contents for the bedroom.

床是卧室空间中视觉的焦点,也是卧室软装的重点。床的软装设计能够为卧室装饰定下情感的基调和主体内容。

It is a typical American style decorative design to use the short columns carved exquisitely as the decorative elements of the four bed corners. The dark brown wood material of the short columns highlights the warmness and gravity of the space.

将床的四脚用雕刻精美的短柱做为装饰的元素,是典型的美式风格的装饰设计。在材料的选用上采用深棕色的木料突出空间的温暖和厚重之感。

The beddings play an important role in the interior design. The leaf-shape patterns on the background colour of green subtly create a consolatory American style.

床被的设计是空间的主要装饰设计，采用绿色为底色，衬托白色的树叶形状图案，比较讨巧的设计将空间营造出不一样的美式风情。

The bed in dark brown, the bright-coloured bed sheet with floral patterns and the gauze curtains integrate gravity and liveliness into the same bedroom, which gives birth to the American pastoral feelings simultaneously.

深棕色的床、鲜艳的花床单以及纱质窗帘，将深沉与活泼统一在同一个卧室空间中，美式的田园风情油然而生。

The carpet and wall cabinet are typically American. Lighting, colours and materials with different textures are integrated into the whole space, completing soft decoration for the bedroom.

空间中的地毯和壁柜凸显出浓厚的美式风情。灯光、色彩以及不同材质的质感相互融合，构成空间的软装风格。

046 SOFT DECORATION: FABRICS IN HOME DESIGN

The wall tiles and the brown timber constitute a typical American style. The bathtub is screened with white curtains, ensuring privacy while being a bit alluring.

壁砖、棕色的木料构成明显的美式风格。用帷幔将浴盆笼罩其内,更增添一丝私密和魅惑。

European Style
欧式风格

European style is one of the most popular home design styles. It puts emphasis on colour and, more often than not, uses marble, textile, and particularly exquisite ornaments such as fine rugs and tapestries to fill spaces with luxuriance and flamboyance, strongly appealing to the eye. The main ornamental elements usually include marble pillar, concave line, fireplace and arch. These elements could bring out a classical European flavour. European style in home design is conceived as a symbol for romance, elegance and quality of living.

欧式风格是住宅的传统装饰风格之一。强调色彩，装修材料常用大理石、织物，多采用精致的装饰品以及地毯和壁挂，使得整个风格显得豪华、绚丽，充满强烈的视觉效果。其主要装饰元素包括：罗马柱、阴角线、壁炉、拱及拱券等，使空间的装饰风格充满欧式古典韵味。住宅采用欧式风格的装饰手段，能够表达空间浪漫、优雅气质和生活的品质感。

In this American-style living room, brownish red is adopted as the basic tune. The cloth sofa and brightly-coloured cushions bring a softly fragrant air to the space.

在棕红色调为背景的美式空间中，用布艺沙发配合鲜艳色彩的靠垫来增加空间的温馨气息。

The typical European-style sofas and the rug perfectly match each other, filling the room with a warm European air. Meanwhile, the fireplace with white pillars goes well with the luxurious chandeliers, making the space feel more sumptuous.

经典欧式造型的沙发与地毯相搭配，烘托出室内温暖的欧式气息。白色的拱柱构成的壁炉与空间内奢华的灯具相搭配，更增添了空间些许的高贵。

The perfect match of the classic pure white European style sofa and carpet composes the traditional European Classicism, while the metal pedestal and the lamppost in unique shapes irradiate a strong modern sense. The integration of tradition and modern makes this kind of mix and match more striking.

经典的纯白色欧式沙发与地毯相配合营造出传统的欧式古典风格。采用金属底座和造型别致的灯柱却焕发出强烈的现代气息。传统与现代相结合，混搭的效果格外引人关注。

Cloth sofa and window curtains are the main soft decoration elements in the space. Correspondingly, European-style wallpaper is adopted to enhance the decoration effect.

空间采用布艺沙发、窗帘等织物作为软装的主要元素,与其相呼应采用具有典型欧式风格的壁纸图案来增添空间的装饰意味。

The luxury of the space is manifested by the details of the refined lamppost and the golden cover of the table with meticulous carving, from which you can feel the quality life.

空间的奢华之处体现在精雕的灯柱和桌子的包金等细节的处理,从精雕细刻里让人体会品质生活。

The engraved golden decoration gives off the air of European luxury. The round table and chairs, along with the chandelier, all create the everlasting classic European style.

镂金装饰凸显欧式奢华，圆形的餐桌和餐椅以及屋顶上的枝型吊灯渲染了空间浓郁的经典欧式气息。

Pale blue is main palette of the American-style dining room. The round wooden dining table, the chandelier hanging above, and the wall lamps create a strong European flavour.

在淡蓝色为主色调的美式餐厅空间中,圆形木质餐桌以及头顶的吊灯、墙面的壁灯凸显出浓郁的欧式格调。

The design of the dining table is the protagonist of the dining room. The square glass table that highlights the spatial gravity and the leather dining chairs that demonstrate the spatial steadiness in collaboration with the furnishings on the table and the modern decorative lighting on the ceiling create a dignified and elegant dining space with the exquisite design.

餐桌的设计是餐厅的主角,玻璃四方桌突出空间的庄重之感,皮质的就餐椅凸显空间的厚重。再搭配桌面的陈设品以及房间顶端的现代灯饰,巧妙的装饰设计营造出一个庄重而又雅致的餐厅空间。

The wooden-framed pictures on the wall, the shining wine glasses on the table, and the milky white leather chairs are all decorative elements for the dining room, highlighting the theme of purity. Contrasting with the brown tabletop, they make the space pleasant for the eye.

墙面的木质挂画、桌上的酒杯以及乳白色的皮质餐椅都是空间的装饰元素，三者突出纯净的主题，与棕色的桌面相对比，空间的装饰意味十足。

Highlighting the elegant flavour of the space, the black and white colour in collaboration with the silver ornaments fulfills the space with intense artistic taste.

黑白两色凸显空间的雅致气息,再配合银质的装饰品,空间散发出浓郁的艺术品位。

The whole space is enveloped in dull colour, among which the black colour in large area creates a lowering and romantic emotional appeal, while the lustre on the surface of the cupboard works in concert with the books piling up on the other side, thus becoming an exquisite decorative element in the space.

空间被暗色调所"笼罩",大面积的黑色营造出低沉、浪漫的情调,尤其是橱柜表面的光泽与空间另一端高高摞起的图书相呼应,成为空间中巧妙的装饰元素。

The whole space in white, especially the brick wall coloured white, radiates the intense modern flavour against its simplicity and plainness, whilst the dining table and the cupboard demonstrate the sophistication and simplicity of modern style design with their clear and distinct edges and corners.

空间以白色为基调,尤其是砖墙也被漆成白色,质朴中散发出浓浓的现代气息。餐台、橱柜则棱角分明,显示现代风格的干练和简约。

The kitchen designed in white looks fresh and bright, which forms a visual contrast with the brown dining table and the solid wood flooring, while it looks cleaner in the embrace of sufficient light.

白色的厨房设计清新、明亮,与棕色的餐桌和实木地板形成了视觉的反差。充足的光线感又将其衬托得更加洁净。

Being consistent with the whole space in colour, those cushions at the head of the bed are an entire part of the bedroom decoration.

摆放在床头的多个靠垫,在色彩的选择上与空间的色彩保持一致,是卧室空间装饰中完整的一个部分。

The heavy curtain adds a sense of dignity to the bedroom whilst the application of the red and yellow colour that echo the colour of the space augments the spatial gentleness and intimacy to some extent, setting off the warmness of the space.

厚重的窗帘增添了卧室空间的深沉，而采用红、黄色彩又弱化了过分的厚重，增添了几分柔和、亲近的感觉，并且与空间的色彩相呼应，烘托出空间的温暖。

Bed is not only the visual focus in the bedroom but the emphasis of the soft decoration of the bedroom. The soft decoration of the bed can set the basic tone of sentiments and the main contents for the bedroom.

床是卧室空间中视觉的焦点,也是卧室软装的重点。床的软装设计能够为卧室装饰定下情感的基调和主体内容。

The warm and soft bed has long been the best place for sleeping. The thick mattress and the design of a large quantity of fabrics in this case endow the bed itself with great dignity and elegance.

温暖、柔软的床始终是人们睡觉的最佳场所，案例中的床采用厚床垫和大量织物的设计，使其散发出雍容华贵的气质。

076　SOFT DECORATION: FABRICS IN HOME DESIGN

The bedroom is designed totally in simple style. Though without any ornaments, the space enjoys more openness and fluency, full of the artistic conception of modern decoration.

卧室空间将简约进行到底，看似没有任何装饰却使空间更加开阔、流畅，充满现代装饰的意境。

With some regional aroma, the well-organized design of the bed, the white bed sheet and the brown cushions, however, bring more modern atmosphere to the bedroom space.

虽然带有一定地域风情,但规整的床体设计、白色的床单和棕色靠垫为卧室空间增添更多现代感。

078　SOFT DECORATION: FABRICS IN HOME DESIGN

The pictures on the wall, the headboard and the skirting in the bedroom are all mounted with golden material. With the crystal chandelier and the light blue colour of the walls, the space is filled with European elegance.

卧室空间中的挂画、床头以及墙角线等采用了镶金的装饰手段,与水晶吊灯以及淡蓝色的空间色彩相呼应,烘托出雅致的欧式风情。

The curtains above the bedhead, the window curtains, the bedding and the carpet are all fabric textures that add a soft and romantic feel to the bedroom, while the bedstands, wall decoration and chandelier are designed with typical European style.

空间中的帷幔、窗帘、床布以及地毯等织物增加空间的柔软和浪漫的感受，而床头柜、墙面的装饰以及水晶吊灯则洋溢着浓重的欧式风格。

082　SOFT DECORATION: FABRICS IN HOME DESIGN

Combining lights, colour and patterns, the decoration of the washroom endows the space with more sense of beauty and visual enjoyment.

洗手间的装饰将灯光、色彩和图案相结合,赋予空间更多的美感和视觉享受。

For bathroom decoration, patterns on the wall and lamps with various shapes are useful tools.

卫生间的装饰风格可以通过墙面图案、灯具造型来表现。

The bathroom in a milk white tune feels warm and soft.

乳白的色调增添了卫生间温馨、柔和的情调。

The design of the half-open washroom where a large amount of light is employed to keep in consistence with the style of the bedroom makes the washroom the transitional area between the bedroom and the toilet.

半开放的洗手间设计，过多采用了灯光将这一区域与卧室的格调保持一致，使洗手间成为卧室与卫生间的过渡。

The collocation of the stream-lined faucet embedded in the mirror glass, the hard texture and the natural patterns of the table facet made of marble and the flowers in faint yellow fills the space with serenity and poetic flavour.

流线型的出水龙头嵌在玻璃镜面当中,台面采用大理石,坚硬的质感、天然的花纹与摆放的淡黄色插花相搭配,空间安静、充满诗意。

The marble table facet in light yellow forms a visual contrast with the white porcelain basin, while the green flowers add some decorative feelings to the space.

鹅黄色大理石台面与白色瓷盆在视觉上形成反差,绿色插花为空间增添些许的装饰感。

SOFT DECORATION: FABRICS IN HOME DESIGN

Marble is adopted in the bathroom on floor and walls, bringing out a sense of strength. Meanwhile, the green plants enliven the otherwise cold and hard space.

空间采用大理石作为地面及墙面的装饰元素，凸显了空间的力度；同时用绿植点缀，为冰冷的空间增添了生机。

The stone material used in large area can enrich the texture of the space, endowing it with more rigidity.

使用大面积的石材能够增加空间的质感,赋予空间更多的刚性。

Colour can enrich the decorative contents of the space. In the space of the toilet, the application of various colours can also create an eye-catching space which leaves men a comfortable feeling.

色彩能够丰富空间的装饰内容,在卫生间空间中运用多种色彩同样可以营造出非常醒目,且让人感觉舒适的空间感受。

Chinese Classical Style
中式经典

The Chinese classical style has its unique form and expression, in terms of layout, colour, texture and design approaches. The typical characteristics of the Chinese classical style design is the use of wood – unique wooden structures are mostly adopted, fully taking advantage of the physical property of the natural material and emphasising the harmony between environment and architecture. In Chinese classical style, painting, carving, calligraphy, craftsmanship and furniture and furnishings are all adopted to create an atmosphere desired. However, it is not simple accumulation of these elements; rather, with your understanding of Chinese culture, you combine modern and traditional elements together, to achieve the aesthetics appealing to the contemporary eye, while carrying forward the tradition into the modern day.

中式经典风格具有独特形式和风格，主要反映在空间布局、色彩、质感和处理手法等方面。 主要特征是以木材为主要建材，充分发挥木材的物理性能，创造出独特的木结构，注重环境与建筑的协调。运用彩画、雕刻、书法和工艺美术、家具陈设等艺术手段来营造意境。中式经典风格不是纯粹的中式元素堆砌，而是通过对传统文化的认识，将现代元素和传统元素结合在一起，以现代人的审美需求来打造富有传统韵味的事物，让传统艺术的脉络传承下去。

The ink-and-wash painting on the wall sets the tone of Chinese style for the space. The wooden chest, as well as the furnishings on it, goes well with the Chinese tone, recalling sweet memories of the past.

水墨画渲染了空间厚重的中式格调，而木质的箱子以及箱子上面的陈设物品都渲染了空间的浓郁的中国风情，能够唤起人们对那些逝去的时光的美好回忆。

The light source is installed under the table top, from which light is projected to set off the round wooden pier hiden under the table, creating the elegant artistic conception. The enlargement of the tearoom in area won't lead to the impairment of its momentum. The lighting design can highlight different spatial emphases.

将光源设置在桌面的下方,灯光从桌下投射出来,将两只隐藏于桌下的圆形木墩凸现出来,营造出幽雅的意境。作为茶室,空间并不会因为面积的增大而气场受到削弱,通过灯光的设计将空间的不同重点得到突出。

In the narrow space, the classic wooden tea table and the round white cushions combine classicalism and modernism. The exquisite woodcarving ornaments on the wall, the quadrangle stool in the corner and the blue and white porcelain vase work in concert with each other, applying more richness, elegance and pleasant flavour to Chinese classical culture.

在狭小的空间中，经典的木质茶台，白色的圆形坐垫，将古典与现代相结合。墙面上的精美木雕装饰与角落的四脚台凳以及青花瓷瓶相互呼应，将中式古典文化渲染得浓郁、雅致怡情。

The tea room in typical Chinese style is used as the home leisure space. Tasting the advanced tea with tea lovers in the room, one can get rid of daily annoyances to enjoy the cosy and unrestrained life.

用充满浓郁中式风格的茶室设计作为家居中的休闲空间。与茶友在其中共品佳茗，将现实中的烦恼抛却，生活变得惬意、潇洒。

096 SOFT DECORATION: FABRICS IN HOME DESIGN

This space is a "mix and match" design of modern and Chinese style. The table, chair and particularly the Chinese chess give off a strong Chinese flavour in the modern space.

这是一个现代中式的混搭设计空间。空间一半的装饰风格是现代的,而通过室内的办公桌椅以及休闲区的棋盘和象棋,又渲染出浓重的传统中式味道。

The design retains the classic Eastern decorative elements such as the palace chair of the Ming and Qing dynasty and employs the modern decorative materials and colours to give modern explanation to the Eastern elements and bring them to full play.

保留了明清时期的太师椅、方桌、屏风等东方经典的装饰元素，在材质上采用现代的装饰材料和色彩，将东方元素进行现代的诠释和发挥。

The classic porcelain vase and the oriental-style light brown screen bring a Chinese air to the bathroom.

空间的粉彩花瓶以及具备经典东方风格图案的茶色屏风为空间增添了几许中式风格的装饰元素。

The shape of the mirror in the bathroom is a typical Chinese pattern – "lucky cloud". It successfully brings a Chinese flavour into the modern space.

卫生间的镜子设计成中式古典的祥云图案，在充满现代气息的整体装饰风格里，将传统中式的韵味发挥得淋漓尽致。

This is a spacious room with a door opening onto the garden and floor-to-ceiling windows that bring the outdoor greenery inside. The classical furnishings in the room, as well as extensive use of wood and stone, all convey the Chinese style.

开阔的空间，一扇门通往院内的花园，宽大的落地窗将室外的绿草尽收眼底。空间内的陈设物品采用古典的装饰造型，配合空间大量的木料及石材的运用，透露些许的中式风格。

A large amount of timber is employed in the soft decoration of the bathroom, which fills men with warmness visually and psychologically so that they can enjoy the happiness while bathing.

洗浴间的软装采用大量的木料,给人在视觉和心理上产生温暖的感受,进而尽情享受沐浴的快乐。

Mix and Match
混搭

As a popular decorative element in interior design, Mix and Match incorporates varied elements, characteristics and styles into one space, pursuing unity among differences; featuring prominence among balances. The mix and match home decoration depends on form rather than being confined to it. Combining the elements from East and West, traditional and modern, classic and fashionable, Mix and Match creates a distinctive space with unique decorative styles.

混搭是室内装饰中非常流行的装饰元素，它将不同的元素、特点、风格于同一空间中，在差异中寻求统一，在平衡中凸显个性。混搭装饰依托于形式却不拘泥于形式。家居中的混搭将东西方元素、传统与现代元素、经典与时尚等装饰元素相互搭配，进而营造出空间与众不同的装饰风格。

The "mix and match" style of the space lies in both colour and decoration. The contrastive colours of white and red bring out a strong visual impact. In decoration, modern and classic styles are seamlessly integrated into one.

空间的混搭之风体现在色彩和装饰风格两个方面。色彩上将乳白与大红相对立,在视觉上能够给人以足够的冲击力。装饰风格将现代与古典融合在一起,混搭的意味浓郁。

In this space, Chinese and European styles are matched together. The whole space is designed with European style, but the furniture – the typical Chinese chairs – reveals some Chinese classical flavour. The mix-and-match effect is eye-catching.

空间采用中式与欧式风格的混搭,空间的整体装饰风格是欧式风格设计,在家具设计方面则增添了带有较强中式风格的木质太师椅,混搭效果十分突出。

With the European glamour, the leather sofa conveys a feeling of seriousness and calmness. Being a perfect match, the tea table is equipped with a metal stand, which enriches the solemn sense. To form a contrast, the designer uses stools in irregular forms to express vivid decorative elements and adopts the stave as the decorative pattern to transmit a relaxed mood. The mix and match of serenity and liveliness, seriousness and easiness creates the decorative effect of the mix and match home decoration.

充满欧式气息的皮质沙发向人们传达出庄重、沉稳之感，与之相配的茶几采用金属支架作为装饰手段，更增添了空间的庄重、严肃。与之相对，设计师用不规则造型的凳子表现活泼的装饰元素，在图案上则采用五线谱作为装饰元素传达放松的情绪。空间将庄重与活泼，严肃与放松相结合，营造出空间混搭的装饰效果。

The perfect match of the classic pure white European style sofa and carpet composes the traditional European Classicism, while the metal pedestal and the lamppost in unique shapes irradiate a strong modern sense. The integration of tradition and modern makes this kind of mix and match more striking.

经典的纯白色欧式沙发与地毯相配合营造出传统的欧式古典风格。采用金属底座和造型别致的灯柱焕发出强烈的现代气息。传统与现代相结合，混搭的效果格外引人关注。

The soft fabric sofa works in concert with the thick leather sofa. To match them, the square wood tea table becomes the spatial focus, rendering the elegant makings inside the limited space.

柔软的布艺沙发和厚实的皮质沙发相互呼应,用木质四方茶几与其相搭配,成为空间的中心,在有限的空间内渲染出高雅的气质。

The modern picture hanging on the wall and the classic copper horse sculpture form the "mix and match" effect of the space.

将具有浓厚现代风格的挂画与代表古典韵味的铜马摆件组合在一起作为空间的装饰物,烘托出空间强烈的混搭装饰效果。

The black leather sofa and the beige fabric settee create the "mix and match" effect. In addition, the wooden box and the metal bowl also form the "mix and match" of furnishings. In this way, the dignified space feels a bit romantic.

黑色的皮质沙发与米色的织物沙发形成混搭，同时木匣与金属盘作为陈设品形成混搭，空间在深沉中凸显些许浪漫的韵味。

The dining table and the dining chairs in the traditional Chinese style and the silver candlesticks and goblet glasses with the strong Western flavour form a sharp contrast to highlight the decorative sense of the space.

空间中的餐桌和就餐椅采用传统的中式风格，银质的烛台和高脚酒杯又带有强烈的西方韵味，采用强烈的、对立的手段来突出空间的装饰意味。

From the dining table to the dining chairs, the simple space design of the dining room conveys the earthy flavour. The ornamental birdcage and the tin cans, however, can express the artistic temperament, enriching the space with more decorative flavour.

简约餐厅的空间设计，无论是餐桌还是餐椅都透露出朴实的气息，作为陈设的鸟笼和锡罐够传达出艺术的气息，为空间增添更多的装饰的味道。

Dining rooms can be used as meeting rooms based on people's needs. In this case, one space has been endowed with two functions by integrating the dining room with the meeting room. There are neither decorations featuring the characteristics of the dining room such as the wine cabinet, the showcase or the square table nor the seriousness of the meeting in that both the lamp and the wall hangings enjoy strong decorative sense.

餐厅也可以成为会议室，关键取决于人的需求。这个空间的案例就是将餐厅与会议室相结合，一个空间两种功能。装饰上既没有突显餐厅的特征——酒架、陈列柜、方桌，也没有强调会议的严肃——台灯和壁饰的装饰感非常强烈。

118 SOFT DECORATION: FABRICS IN HOME DESIGN

The space is made warm and elegant by the decorations such as the soft carpet, the comfortable sofa and pedal, the tea table made of assembling battens and the semicircular window that brings more brightness into the interior space.

半圆形的窗户使室内空间格外明亮，柔软的地毯，舒适的沙发、脚踏，用木条拼搭成的小几，空间温暖而又雅致。

SOFT DECORATION: FABRICS IN HOME DESIGN

The office is decorated in a simple way with paintings, lighting and furnishings. The quiet atmosphere reveals the artistic taste of the owner.

简约的办公空间采用简单的装饰手段：挂画、灯光以及陈设品，安静中隐含着主人的艺术品位。

The office design with thorough modern feeling adopts the milk white colour to highlight the beauty of simplicity.

现代感十足的办公设计,采用乳白色凸显简约之美。

The delicate decoration can not be applied to all the elegant space. For most ordinary families, perhaps a simple ornament can highlight the spatial elegance just like the statue in the corner in the picture above.

并不是所有的雅致空间都需要有精致的装饰内容。对于大多数的普通家庭而言，或许一个简单的摆设就能够凸显空间的雅致感。上图中角落里的雕像就是如此。

SOFT DECORATION: FABRICS IN HOME DESIGN

The two blue and white porcelain jars make the European-style space feel a bit different at once, injecting a strong Chinese flavour.

空间的风格因为桌上的两只高大的将军罐和罐上所描绘的代表经典中式风格的青花,在强烈的欧式风格空间中渲染出浓郁的中式风格韵味。

The vestibule design of the space features a display wall on which the frames painted in white, as exhibiting subjects, are designed in typical modern style.

此空间的玄关设计以展示墙的风格为主，用漆成白色的画框作为其展示的主体，设计的概念极具现代派。

128 SOFT DECORATION: FABRICS IN HOME DESIGN

The vestibule design here creates a serene and peaceful space with the help of light, green plants and some small decorative articles.

此处的玄关设计,采用灯光、绿植和小摆件来营造一处祥和、安宁的角落。

To coordinate the brown colour of the wooden cabinet, the space is decorated in light yellow which can highlight the gentleness and warmness of the space. The design of the rooftop is different from the traditional kitchen in that it exerts commendable decorative effect while increasing the space height. Moreover, the green plants brought into the kitchen can not only purify the air but act as an attractive view.

配合棕色的木质橱柜，空间的色彩采用鹅黄，突显出空间的柔软和温暖，顶棚的设计区别于传统的厨房，在增加空间高度的同时，也起到了很好的装饰效果。除此之外，将绿植引入厨房，既可以净化空气，又是其中的一道美丽风景。

The kitchen design in traditional style can bring out intense classical beauty. Even inside the kitchen, you can feel the elegant spatial ambience of European Classicalism.

传统风格的厨房设计能够散发出浓郁的古典美,即使身处厨房,你也能够感受到欧式古典那优雅的空间气息。

The light in warm colour makes the space bright and full of tender feelings against the background of the sharp contrast between purplish red and milk white and the collocation of the natural light and the lamp light.

空间采用枣红色与乳白色相对，配以暖色的灯光，空间在自然光与灯光的烘托下，亮丽且充满温情。

The strong liveliness set of against the space in blue and green colour leads men to feel the briskness and gorgeousness.

空间采用蓝绿相间的色调渲染出强烈的明快感,让人能够感受到清爽和明艳。

The texture of the material is the most natural decorative expression. All these elements such as the solidness of the metal, the tenderness of the timber and the strips and texture of the marble table facet can leave people distinct visual and tactile feelings.

材质的肌理是最自然的装饰表达，金属的坚硬与木料的温和，理石台面的条纹和质感给人的视觉和触觉感受都是十分明显的。

The modern style kitchen design, perfectly integrating the ingenious conception with the modern decorative material, expresses the glamour of modern style through the shape and the texture of material.

现代风格的厨房设计将巧妙的构思与现代装饰材料完美地结合，采用造型、材料的质感来表达现代风格的魅力。

Forming a sharp contrast in vision, the white cushion and the deep black headboard set off by the warm light create a pure and serene rest space.

白色的靠垫与深黑色床头形成鲜明的视觉反差,在温暖灯光的衬托下,营造一个纯净、安详的休息空间。

The Chinese-style wooden tea tray and cups, and the European-style portray hanging on the wall form the "mix and match" effect in the bedroom.

木制的茶盘、茶杯代表着中式风格,墙面上的挂画又透露出欧式的气息,空间的混搭风格由此越发浓烈。

With some regional aroma, the well-organized design of the bed, the white bed sheet and the brown cushions, however, bring more modern atmosphere to the bedroom space.

虽然带有一定地域风情,但规整的床体设计、白色的床单和棕色靠垫为卧室空间增添更多现代感。

Timber, stone, textile and plant are blended together in the bedroom, creating a "mix and match" effect with different textures.

这间卧室空间将木料、石料的材料肌理与织物、绿植元素杂糅在一起,进而起到混搭的装饰效果。

The small floor lamp in simple style standing next to the curtain gives out soft and warm light against the curtain, filling the space with artistic conception.

造型简约的小地灯紧挨着窗帘摆放,灯光映衬在窗帘上,柔和、温暖而又充满意境。

The orchids at the bed head set off by the dim light from the bedside wall lamp, look so elegant that they become the major ornaments in the soft decoration of the space.

床头的壁灯投射出淡淡的光亮,将床头的兰花衬托出高雅的气质,成为空间中软装饰的重要点缀。

The walls are clad with wooden panels, effectively creating a warm ambience in the bathroom.

通过墙面装饰以及增加木质材料的比例，增加了卫生间的温馨气息。

With the ability of enriching the space texture and increasing the stern atmosphere, the Mosaics set off by the light from the small white pendant lamp endow the space with simplicity, purity and solemnity.

马赛克能够增加空间的质感和冷峻的气息,在白色的小吊灯的映衬下,空间显得简洁而又严肃。

Soft Decoration in Home Space
家居软装饰

Life is inseparable from space. In home space, it is difficult to give a separate definition to life space. To make a distinction between the characteristics and properties of different spaces, living space in this book refers to the living room and the dining room. Living rooms and dining rooms serve as the centre of family life and the main sites for family gathering. Compared with other home spaces, living space enjoys the most intimate relationship to men, thus having more meticulous and rigorous requirements for soft decoration.

Living room is the main room of family life, the locality for such activities as sitting, reading, relaxing and gathering as well as the space to receive and entertain guests. What deserves the designers' studious consideration more before the soft decoration design of the living room is the spatial role of the living room: whether it is a place mainly for resting or working; a site for guests entertaining or receiving; a space featuring the general decoration or the collection display. However, the role of the space is determined on the basis of men, their demands and tastes. In the soft decoration design of the living room, the design content of different parts including doors, windows, the sofa and the like should be determined in line with the frame feature of the space. Moreover, the diversified effects of the decoration viewing from each and every angle should be also took into consideration. The design of the living room would exert more multidimensional effects and still smoother and more natural proceeding with the above aspects.

Dining room is a place where we can have dinner, talk and share happiness as well as a locale to get together all the family members and enjoy the good times. Located between the kitchen and the living room, the dining room seldom enjoys its own independent space. For the non-independent dining room, the one next to the kitchen or the living room, the soft decoration would begin with bringing changes into the design to make the dining room a functional space rather than simply regarding it as the only place for dining. Meanwhile, dining room can be used to have a family meeting as well and the dining table acts as the meeting holder besides the dishes carrier.

As the core of home soft decoration, the design of the rest space should manifest the personality and style of the owner, which needs the designer to set the basic tone and the decorative style of the space in the decorative design process according to the owner's demand. Apart from offering people the place for sleeping, the bedroom can also perform such functions as changing room, powder room, study and media room, which needs us to attach great importance to the style of soft decoration while donating various functions to it.

In home space, almost all the spaces such as the dining room, the living room and even the bedroom where a couple sleep in the hug of each other are open spaces that can be shared together. Only the lavatory that needs to be locked is the space enjoyed by somebody himself. In this sense, we call lavatory the private space. The soft decoration of the lavatory, the single private space in the home space, should take into account its functionality. The lavatory should comprise such functional areas as the washroom, the bathroom and the toilet. The soft decoration of different areas should be carried out on the basis of the varied functions of each part so as to delight people's vision and relax their mind.

生活离不开空间，在家居空间中，如果单独定义生活空间似乎比较困难，但为了区别不同空间的特点和性质，本书将起居室和餐厅作为生活空间。起居室和餐厅是家庭生活的中心，也是家庭成员相聚在一起的主要场所。相对于其他家居空间，生活空间与人的关系最为密切，对于软装饰的要求也更加精细和苛刻。

起居室是家庭生活的主要场所，是坐、读、放松、聚会等活动的发生地，同时也是会客的空间。在进行软装设计之前，需要努力去思考的是起居室的空间角色——是以休息为主还是以工作性质为主；是以娱乐为主还是以会客为主；是以普通的装饰为主还是成为陈列收藏品的空间。起居室的软装设计，要根据空间的建筑框架特征来确定不同位置的设计内容——门、窗、沙发以及其他。另外还需要考虑从不同的角度来看起居室的视觉效果会是怎样。从以上几个方面入手来进行空间软装设计就会使起居室的设计更具多维度的效果而且更加自然、流畅。

餐厅是我们吃饭、交谈、分享快乐的空间，同时是将家庭成员聚集在一起，享受美好生活时光的场所。家居中的餐厅通常是处在厨房与起居室之间的位置，很少有单独开辟的独立空间。对于更多的非独立的餐厅空间，或者是紧挨着厨房或者是紧挨着起居室，这样的餐厅软装饰可以不拘泥于将其仅仅作为就餐的空间，将变化带入设计，使餐厅成为多功能的空间，餐厅同时可以作为家庭会议的空间，而餐桌除了用来吃饭之外，还扮演着承载会议的角色。

卧室的设计要体现出房间主人的个性和风格，这就要求在装饰设计中应该根据房间主人的需求来确定空间的基调和装饰风格。卧室除了为人们提供睡觉场所之外，通常还要扮演着其他角色：更衣室、化妆室、书房、电视房等等，这就需要在赋予卧室不同功能的同时兼顾到其软装的风格。

在家居空间中，餐厅、起居室等空间都是开放的、多人共享的空间，甚至是卧室也可以两个人相拥而眠，只有卫生间是需要锁门的、只允许一个人独自享用的空间，因此，我们将卫生间称为私密空间。作为家居中唯一的私密空间，卫生间的软装设计更多要考虑其功能性的体现。卫生间通常必须包括的几个功能区域有：洗手间、洗浴间、厕所。不同区域的软装设计要结合区域本身的功能特点合理安排，以愉悦视觉，放松心情为主要目的。

Living Room
起居室

There are many elements included in the soft decoration design of the living room such as furniture, fabric, work of art, wood, stone, metal and glasswork, the kinds of materials usually used. Various designs with the diversified combinations of these elements can bring about different decorative contents and spatial effects. The design of the living room can proceed with the tea table or the coffee table by placing a set of sofas and chairs around the room and using display stands, bookcase or artwork exhibit as part of the wall space design. This kind of design is familiar to us with the advantage of creating the comfortable ambience. Besides that, sofa can also be used as a focus of the space design while employing lights and art works as means of rendering the atmosphere of space so as to devise the living room as an elegant space for spiritual relaxation and enjoyment. In terms of the ones in large areas, they can be divided into varied zones, for instance, the life-rest area and the life-work area. Either way, the glamour and sentiment of the space and the individuality of each area are indispensable. The soft decoration of the living room should begin with the location of different areas and then the design of lights, fabric and other details. Light is both the major element affecting the sense of sight and moreover the atmosphere conditioner of the interior space, an important means of rendering sentiments. Fabric is the gentlest strength among all the decorative elements, endowing people with a gentle and warm feeling. Fabrics are indispensible in the living room. Without fabrics, a living room would be somewhat uncomfortable. Therefore, cushions, carpets, tablecloths and the like are commonly seen in living room design. Soft decoration details usually embody a good design, or an attitude towards design.

起居室的软装饰设计包含的元素有很多，通常都能够用到家具、织物、艺术品、木材、石料、金属、玻璃制品等。这些元素的组合有多种方式，不同的设计所表现的内容和营造的空间效果也是不同的。起居室的设计，可以以茶几或咖啡桌为设计的起点，将一组沙发或者椅子环绕周围，墙面的设计可以采取陈列架、书柜或者艺术品展示等方式。这种设计方式是比较常见的，优点是能够营造舒适的空间氛围。此外还可以将沙发作为空间设计的中心，用灯光和艺术品作为渲染和烘托空间气氛的手段，将起居室设计成为精神放松、精神享受的高雅空间。对于空间较大的起居室，可以将其分割成为不同的区域，比如生活—休息区和生活—工作区。而无论以哪种方式来设计，空间的魅力、感情和各部分的个性同样是不可或缺的。

起居室的软装饰，在确定了各部分的位置之后，需要考虑的是灯光、织物以及细节等方面的设计。灯光是影响到人们视觉的主要元素，同时灯光也是室内空间的调情师，是渲染情调的重要手段。织物是装饰元素中最柔软的力量，能够给人以柔和、温暖的感受。起居室中如果缺少了织物，就会给人感觉很不舒服，因此，靠垫、地毯、桌布等织物的设计都是必不可少的。细节的处理往往能够于小处见大智慧，同时也是体现设计精神的手段。

The large carpet, cloth sofas and floor-to-ceiling window curtains make the living room feel warm and noble.

大面积的地毯、布艺沙发、落地窗帘等织物的装饰设计将起居室的格调渲染得高贵而又祥和。

In the living room, European and American styles are mixed together. With the warm light, the space feels quiet and comfortable.

起居室的装饰融合了欧式风格和美式风格的特点,在淡淡的灯光衬托下,空间显得宁静、舒适。

Metal decoration is extensively adopted in the living room, such as the table, the lamp standards and the exquisite metal piece at the fireplace. The texture and colour of the material make the space feel dignified and luxurious.

起居室运用了大量的金属色,桌腿、灯柱以及壁炉处的金属装饰品,将空间烘托出厚重而又奢华的装饰效果。

Simplicity is a kind of life wisdom, which means to create comfortable, orderly and pleasant elements in simple methods. The vogue connected with the simple decorative style is a popular trend at present and also the manifestation of the respect for freedom. The application of simple and fashionable embellishment in the soft decoration of the living room can produce a strong sense of modern life.

简约是生活中的智慧,用简单的表现手段来营造舒适、条理和快乐的元素。与简约装饰风格相接近的时尚,是在当前所流行的风潮,尊重自由的表现。简约时尚的装饰体现于起居室的软装饰中,能够营造出浓郁的现代生活气息。

The blue cloth sofas create a calm and cool ambience. With the modern painting hanging on the wall, the living room feels simple yet elegant.

蓝色的布艺沙发为空间增添一丝冷静的气氛，配合墙面具有强烈现代感的装饰画，整个起居室的设计凸显简洁的味道。

Radiating a gentle and warm flavour, fabric sofas make people feel intimate. Being the opposite, the wall conveys a sense of crudeness and a cold visual effect. The space is designed in the mix and match style combining the cold and warm decorative effects.

布艺沙发传达出柔软、温暖的气息，令人感到亲近。与之相对立，水泥墙面却透露出一丝生硬感和冰冷的视觉效果。空间将冷、暖装饰效果形成混搭装饰风格。

Colour is the main decoration approach in the design of the living room. Various bright colours visually enrich and invigorate the limited space.

色彩是这个起居室的主要装饰元素，多种艳丽的色彩相互呼应，将空间渲染得富有动感，有限的空间显得内容愈加丰富。

Colour is a key decoration in living room design. In dark-tuned spaces, bright yellow or blue is adopted and easily becomes the focal point.

色彩是起居室的主要装饰元素,在暗色调的空间中用鲜艳的黄色或者蓝色作为装饰色,很容易成为人们的视觉焦点。

160 *SOFT DECORATION: FABRICS IN HOME DESIGN*

Dining Room
餐厅

In the soft decoration design of the dining room, the top priority of consideration goes to the functional partition of different spatial parts such as the location, shape and characteristic of the dining table, the quantity, sizes and enclosure effect of the chairs—if they look crowed after the enclosure, the spatial location of the wine cabinet and the showcase, the arrangement of lights and the placement of artworks.

The design of the dining room is inseparable from the functional elements which refer to many other functions besides the main one—the need for dinning, for instance, the dining room can be the locale for such activities as family gathering, family activities, making telephone calls and receiving mails. In terms of the mere dining function, the soft design of the dining room should take into account the varied dining contents for breakfast, lunch and supper, people's different moods and physiological needs.

There are many kinds of decorative items, namely, linen tablecloth, napkin, silk ribbon, porcelain, glassware, pewter, silverware, dinnerware, candlesticks, flowers, vases, lacquers, coral and cobblestones, which are commonly seen. The comprehensive application of these elements can augment the spatial atmosphere of soft decoration and set off a strong sentiment.

Moreover, the soft decoration of dining rooms needs to give consideration to the rest four elements including art, lighting, display and comfortableness.

We all have a longing for art. In dining rooms, such a longing is usually embodied in pictures on the wall, artistic ornaments, etc. Lighting is the soul of decoration. Lighting in a dining room is particularly important in creating a soft, cosy atmosphere. In dining rooms we often have cupboards for wine collection and tableware storage, or even for artistic display, visually enriching a dining space. Art, lighting and cupboards design should all be combined to make a dining room feel comfortable. After all, a good dining room design is to offer an environment where you can have a pleasing dining experience.

餐厅的软装饰设计,首先要考虑的是划分空间的各部分功能区域:餐桌的位置、形状和特点;椅子的数量、大小以及围合后是否显得拥挤;酒柜、陈列柜等所处的空间位置、灯光的安排、艺术品的放置等因素,都需要在这一阶段加以考虑。

餐厅的设计离不开功能因素,除了需要考虑主要功能——就餐的需要之外,还要考虑更多其他的功能,比如餐厅也可以是家庭聚会、家庭作业、打电话、收发邮件等活动的场所。单纯就餐功能而言,餐厅的软装设计也应该照顾到早餐、中餐、晚餐等不同就餐时段的内容和人们的不同情绪以及生理需求。

餐厅的装饰物品包含很多元素,比较常见的如:亚麻桌布、餐巾、丝带、瓷器、玻璃制品、锡器、银器、餐具、烛台、花、花瓶、漆器、珊瑚、鹅卵石等。这些元素的综合运用,可以增加餐厅空间的软装气氛,烘托出强烈的情绪感。

此外,餐厅的软装饰设计还需要考虑到其他4个维度:艺术、灯光、陈列和舒适。

艺术是每个人都向往的,餐厅中的艺术可以通过挂画、艺术装饰品等来表现。灯光是装饰的灵魂,餐厅的灯光设计要柔和、温暖,能够起到烘托气氛的作用。陈列通常指餐厅中的酒柜、餐具储藏柜或者是专门的艺术品陈列柜等家具的陈列设计,陈列设计能够增添空间的气势,使空间显得饱满、富有张力。舒适是整个空间在考虑到以上3个方面的因素之后,空间整体体现出的让人很舒服、不排斥的心理感受。

Simple and convenient, the modern style dining room design is closer to the real life.

具备现代风格的餐厅设计,简约、方便,更加贴近人们的现实生活。

Soft decoration in the dining room is realised with the glass candlesticks and the painting hanging on the wall. There seem to be many elements in the space, but it won't feel too much complicated. Symmetry plays an important role.

餐厅的软装通过桌面上的玻璃烛台和插花来表现，看起来元素较多，但并不复杂，采用对称的方式将软装表现得很到位。

The candelabrum on the dining table is the focal point in the space. The vivid green candelabrum looks like a living bamboo, effectively enlivening the dining room.

绿色的烛台是空间装饰的焦点,运用夸张的色彩设计将鲜艳的翠绿色作为烛台的装饰元素,仿佛在餐桌上矗立的具备旺盛生命力的翠竹,为餐厅空间增添了一丝生动的气息。

The colour of yellow sets the warm tone of the dining room, while the green wine glasses and chandelier make the space feel a bit calm and cool.

将棕色作为餐厅的主色调烘托出空间中的暖意,绿色的玻璃酒杯以及吊灯在暖色调的衬托下又赋予了空间冷静的品格。

170 SOFT DECORATION: FABRICS IN HOME DESIGN

Bedroom
卧室

The design of the bedroom, the rest area in the home space, should attach much importance to two thematic elements: peace and safety. The sense of safety needs to be created by the application of more soft fabrics in the bedroom space such as curtain, bedclothes and carpet, which are the protagonists of the soft decoration for the bedroom. People's feeling and perception of safety begin with the tactile sense. The different textures of the materials such as silk, cotton cloth and lingerie leave people varied tactile impression, visual perception and degrees of consolation. Fabrics with different material qualities used as decorative elements can exert various decorative effects.

Besides fabrics, furniture undoubtedly serves as the principal part of the bedroom space, in which the bed attracts people's most attention among all the furniture as they can sleep, read, relax or even eat and watch TV in bed. The decorative design of bed, as the focal point of decoration of the bedroom space, comprises the decoration of bed head and that of the body.

Lighting, as the significant decorative element that cannot be ignored, is the main device to create the spatial sentiment. The lighting design should be able to create varied atmosphere in line with different demands. It would be best that the intensity of light can be controllable so as to satisfy people's different psychological needs. Being a part of the decorative design of the bedroom, the floor decorations including the solid wood floor and the carpet can present people with natural and warm feelings.

卧室是家居中的休息空间，和平、安全是卧室装饰设计的主题元素。营造安全感需要在卧室空间运用更多的织物：窗帘、床单、地毯等，这些柔软的织物是卧室软装的主角。人们对于安全感的体会和感受是从触觉开始的，丝绸、棉布、亚麻制品等不同的质感其触感是不同的，给人的视觉感受和心里慰藉程度也有所不同。用不同材质的织物作为软装元素可以体现出不同的装饰效果。

除了织物，家具无疑是卧室空间的主体，而床又是卧室家具里人们第一眼的焦点，人们可以在床上睡觉、读书、休闲，甚至吃东西、看电视。床的装饰设计是卧室空间装饰的重点，床的装饰又分为床头装饰和床身装饰。

灯光是卧室设计中不能忽视的装饰元素，它是渲染空间情调的主要手段。卧室的灯光设计要根据不同的需求渲染不同的情调，光亮的明暗度最好是可以调节的，以满足人们的不同心理需求。地板同样也是卧室中装饰设计的一部分，采用实木地板或者地毯做装饰，能够给人自然、温暖的感受。

174 SOFT DECORATION: FABRICS IN HOME DESIGN

SOFT DECORATION: FABRICS IN HOME DESIGN

Cushions are the fabrics that are commonly seen in the bedroom. Cushions can not only be enjoyed in people's arms, on their backs or under their heads, but become the key decorative element of the space by means of soft decoration.

靠垫是卧室空间中比较常见的织物。靠垫不仅可以供人们抱在怀里、靠在后背或是枕在头下，而且通过软装设计也能够成为空间重要的装饰元素。

The bedroom feels warm and cosy with soft decoration such as the carpet, curtains and fabric furniture. The curtains, the bed sheet and the cushion on the sofa with the same pattern set a unified pattern scheme for the room.

卧室空间被地毯、窗帘、布艺家具等软装元素烘托出浓郁的温暖气息，窗帘与床布、沙发靠垫的花色图案采用同一种样式，使空间的软装成为统一的整体。

Washroom
洗手间

Such activities as daily washing, personal cleanness after using the toilet and clothes sorting are all completed in the washroom, so it is an area with the highest utilization rate in the lavatory. The design elements of the washroom usually comprise the washbasin or the wash dais, the locker, the storage rack, the mirror, the light and the like.

The washbasin and the wash dais are inseparable from each other. The washbasin is usually made from ceramic whose exquisite texture can leave men the comfortable sensation both visually and tactually. It is certain that different materials can achieve different decorative effects which lead to varied psychological feelings of people. The wash dais refers to the table facet used to support the washbasin and the cabinet under the table facet. The choices and design of different table facets, the main decorative elements of the washrooms, may produce different visual perceptions as well.

The locker and the storage rack are used for storing articles. The accessories with strong decorative sense and the articles for daily uses which can produce decorative effects can enliven the spatial atmosphere and intensify the decorative effects after being placed on the storage rack.

As the necessities in the washroom, mirrors are used by people to watch themselves clearly. The decorative function of the mirror, an decorative element that cannot be ignored, lies in that it can enlarge the space visually by enriching the space content with its reflective function. The lighting design can enliven the spatial atmosphere. The lighting of the washroom should feature brightness and it is necessary to offer the local lighting.

洗手间是卫生间中利用率最高的一个区域，人们平时的洗漱、如厕后的个人清洁、洗澡之后整理衣着等活动都要在这一区域中完成。洗手间的设计元素通常包括：洗手盆及洗手台、储物柜、储物架、镜子、灯光等。

洗手盆和洗手台是不可分离的整体，洗手盆通常用陶瓷以其细腻的质感给人以视觉和触觉的舒适感。当然，不同的材质能够营造不同的装饰效果，对人们产生不同的心理感受。洗手台是支撑洗手盆的台面以及台面之下的柜子。通常是洗手间的主要装饰元素，不同台面的选择和设计同样会产生不同的视觉感受。

储物柜和储物架的功能是存放物品，将装饰感较强的小饰品或带有装饰效果的用品放在储物架上，可以调节空间的气氛，增强空间的装饰效果。镜子是洗手间的必须品，人们需要对着镜子才能够看清自己。镜子对空间的装饰在于其能够增加空间的视觉体量，通过反射效果丰富空间的内容，是不容忽视的装饰元素。灯光的设计能够调节空间的气氛。洗手间的灯光要以明亮为主，尤其是提供局部的照明是非常必要的。

Combining light, colour and patterns, the decoration of the washroom endows the space with more sense of beauty and visual enjoyment.

洗手间的装饰将灯光、色彩和图案相结合,赋予空间更多的美感和视觉享受。

The combination of hard stone and lively green plants creates the contrastive juxtaposition of coolness and warmth. In this way, the bathroom is harmoniously enriched.

一面是坚硬的石料作为主要的装饰元素凸显出冷峻的效果,另一面用几束绿植增添空间活泼的气息,使洗手间的装饰在变化中达到和谐。

Entertainment and Leisure
娱乐与休闲

Entertainment and leisure are the indispensable components of modern life. People have long been yearning for the recreational life style. With the further development of New Hedonism, people suggest more need for entertainment. Recreational life is inseparable from the entertainment space, while the public entertainment space is too noisy. In this sense, the home entertainment space can further highlight the understatement and the spiritual pursuit, showcasing its unique taste.

娱乐与休闲，是现代生活中不可缺少的组成部分，生活娱乐化是人们所向往和追求的，随着新享乐主义的生活理念不断深入，人们对于娱乐的需求也愈加强烈。娱乐生活离不开娱乐的空间，公共场所的娱乐空间又太过于喧闹，因此家庭娱乐空间则更加突出内敛和精神追求，更显独特的品位。

Soft decoration can make the recreational area full of modern atmosphere or endow it with the elegant temperament possessed by calligraphy and painting. No matter what kind of soft decoration, it needs people's perception and understanding of entertainment from their inner hearts.

可以通过软装将其设计成充满现代气息，也可以将其赋予"琴棋书画"的优雅气质。无论哪种手段的软装设计，都需要人们心底对于娱乐的解读和理解。

A chaise lounge is set by the window. Without too much decoration, the space feels cosy and relaxing. It seems simple but reveals wisdom of the designer.

临窗的一张小躺椅,摒弃了过多的修饰进而凸显放松、舒适的格调,简约中凸显设计的智慧。

SOFT DECORATION: FABRICS IN HOME DESIGN

An entire wall is filled with bookshelves on which there are a large number of books. Meanwhile, the red sofas present men a vibrant visual enjoyment in colour, further highlighting the intense cultural atmosphere in the space.

房间的一面墙都被书架所填充，书架上装满了大量的图书，红色的沙发在色彩上给人跳跃的视觉感受，而且更加着重地凸显了空间的浓郁的文化氛围。

The American style office space combines the pastoral style with the traditional European classical decorative style, creating the dignified and lively spatial ambience. The soft decoration design which can make people feel the serious atmosphere and become easy at the same time is suitable for the home office space.

美式风格的办公空间,将田园风情与传统的欧式经典装饰风格相结合,能够营造出厚重而又活泼的空间气氛,对于办公空间而言,能够让人既体会到严肃的气氛,但又相对放松,更适合家居空间的办公软装设计。

Soft Decoration Details
软装细节

A good interior designer pays attention to every detail in a space, especially soft decoration details, such as lighting in a corner and the position of a piece of furnishings. Soft decoration in home spaces should go well with the lifestyle of the owner of the house because the details of soft decoration define the details of his life. Proper soft decoration details would help make his life easier and enjoyable.

Details are not necessarily complicated. On the contrary, simple details would reveal wisdom of the designer. Particularly in modern time, we pursue simple home spaces with few elements to create relaxing atmospheres. Maybe a detail is not eye-catching, but, if designed with ingenuity, it would bring some surprise to your daily life. That's what we call detail.

A detail may be the finishing touch that brings a space to life. For example, a green plant in a bathroom would be the focal point of the boring space. In other cases, details could help set off a desired atmosphere. They can be some pieces of artwork, or even several books. You never pay attention to them seriously, but they are really indispensable to the ornamentation of the space.

好的设计要注重过程中的每一个细节，在家居软装饰中的细节体现在方方面面，可以是一抹灯光的处理，也可以是一个陈设品的摆放位置。家居软装设计要密切结合人们的生活方式，软装细节即是生活中的细节，将软装细节处理恰当会让生活更加轻松和惬意。

细节的处理不一定要很繁琐，相反越是简单的设计越能够彰显智慧，尤其是现代生活中更要求空间的简约美，不需要用太多的元素就能够使空间凝练出轻松的氛围。很多细节未必是起眼的，但在设计过程中如果独具匠心，会使人在空间生活的过程中不断带来小惊喜。这就是细节。

细节在空间中的作用，有起点睛作用的，比如浴室里的一盆绿植，这些细节设计常常会成为整个空间装饰的聚焦点，用来吸引人的目光和注意力。也有起烘托作用的，比如桌面上的陈设物品，可以是艺术品，也可以是几本普通的书。这些装饰细节人们一般不会去看它，但是它却是空间装饰的一个部分，去掉就会让人感觉很别扭。

Furnishings

陈设

Furnishings can be categorised into two types: functional and decorative. Functional furnishings have both utility value and appreciation value, such as furniture and lamps. For decorative furnishings, the emphasis is put on appreciation value, such as art work and souvenirs. Furnishings are important decorative elements in interior design, particularly for soft decoration.

In home spaces, furnishings are not only decorative elements; more importantly, they have cultural and artistic value. Furniture, as the main constituent, has been developing from functional pieces to artistic works and has become an important way of decoration in home design. For decorative furnishings, ostentation or extravagance is not the point; nor does it involve too much technical matter. In fact, in the design of furnishings, emphasis should be laid on the relationship between furnishings and space, in terms of functionality, scale, style, etc. Function and artistic appreciation should achieve a harmonious balance.

陈设包括功能性陈设和装饰性陈设。功能性陈设指具有一定实用价值又兼具观赏性的陈设，比如家具、灯具等。装饰性陈设指以观赏性为主的陈设，如工艺品、纪念品等。陈设是空间的装饰元素，也是重要的软装饰细节。

住宅装饰设计中的陈设，不仅仅作为装饰品而具有观赏价值，更为重要的是它还具备一定的文化价值和艺术气息。家具是陈设装饰艺术中的主要构成部分，随着时代的不断进步，其艺术性越来越受到人们的重视，已经成为现代家居装饰设计中的主流装饰手段。装饰性陈设品通常被称为摆设，其在空间的装饰设计既没有必要片面追求排场又不要过分强调技术性和科学性，设计的重点应该放在陈设与空间的功能、比例关系以及风格之间的关系，强调空间内部功能与艺术效果之间的统一。

Furnishings
陈设

Furnishings can be categorised into two types: functional and decorative. Functional furnishings have both utility value and appreciation value, such as furniture and lamps. For decorative furnishings, the emphasis is put on appreciation value, such as art work and souvenirs. Furnishings are important decorative elements in interior design, particularly for soft decoration.

In home spaces, furnishings are not only decorative elements; more importantly, they have cultural and artistic value. Furniture, as the main constituent, has been developing from functional pieces to artistic works and has become an important way of decoration in home design. For decorative furnishings, ostentation or extravagance is not the point; nor does it involve too much technical matter. In fact, in the design of furnishings, emphasis should be laid on the relationship between furnishings and space, in terms of functionality, scale, style, etc. Function and artistic appreciation should achieve a harmonious balance.

陈设包括功能性陈设和装饰性陈设。功能性陈设指具有一定实用价值又兼具观赏性的陈设，比如家具、灯具等。装饰性陈设指以观赏性为主的陈设，如工艺品、纪念品等。陈设是空间的装饰元素，也是重要的软装饰细节。

住宅装饰设计中的陈设，不仅仅作为装饰品而具有观赏价值，更为重要的是它还具备一定的文化价值和艺术气息。家具是陈设装饰艺术中的主要构成部分，随着时代的不断进步，其艺术性越来越受到人们的重视，已经成为现代家居装饰设计中的主流装饰手段。装饰性陈设品通常被称为摆设，其在空间的装饰设计既没有必要片面追求排场又不要过分强调技术性和科学性，设计的重点应该放在陈设与空间的功能、比例关系以及风格之间的关系，强调空间内部功能与艺术效果之间的统一。

Soft Decoration Details
软装细节

A good interior designer pays attention to every detail in a space, especially soft decoration details, such as lighting in a corner and the position of a piece of furnishings. Soft decoration in home spaces should go well with the lifestyle of the owner of the house because the details of soft decoration define the details of his life. Proper soft decoration details would help make his life easier and enjoyable.

Details are not necessarily complicated. On the contrary, simple details would reveal wisdom of the designer. Particularly in modern time, we pursue simple home spaces with few elements to create relaxing atmospheres. Maybe a detail is not eye-catching, but, if designed with ingenuity, it would bring some surprise to your daily life. That's what we call detail.

A detail may be the finishing touch that brings a space to life. For example, a green plant in a bathroom would be the focal point of the boring space. In other cases, details could help set off a desired atmosphere. They can be some pieces of artwork, or even several books. You never pay attention to them seriously, but they are really indispensable to the ornamentation of the space.

好的设计要注重过程中的每一个细节，在家居软装饰中的细节体现在方方面面，可以是一抹灯光的处理，也可以是一个陈设品的摆放位置。家居软装设计要密切结合人们的生活方式，软装细节即是生活中的细节，将软装细节处理恰当会让生活更加轻松和惬意。

细节的处理不一定要很繁琐，相反越是简单的设计越能够彰显智慧，尤其是现代生活中更要求空间的简约美，不需要用太多的元素就能够使空间凝练出轻松的氛围。很多细节未必是起眼的，但在设计过程中如果独具匠心，会使人在空间生活的过程中不断带来小惊喜。这就是细节。

细节在空间中的作用，有起点睛作用的，比如浴室里的一盆绿植，这些细节设计常常会成为整个空间装饰的聚焦点，用来吸引人的目光和注意力。也有起烘托作用的，比如桌面上的陈设物品，可以是艺术品，也可以是几本普通的书。这些装饰细节人们一般不会去看它，但是它却是空间装饰的一个部分，去掉就会让人感觉很别扭。

In a corner of the dining room, exhibition and storage functions are integrated together. The application of the brownish red timber makes it work in harmony with the total colours. The American style detail treatment embodies the sedate and elegant style.

将陈列与储藏功能相结合，采用棕红色的木质材料，与空间整体的色调保持一致，美式的细节处理体现出沉稳、高雅的风格，为空间的软装增色不少。

198 SOFT DECORATION: FABRICS IN HOME DESIGN

200　SOFT DECORATION: FABRICS IN HOME DESIGN

The space features the modern simple decorative style highlighted by the simple but elegant colours. The space decoration is characterized by the design of the pendant lamp and the ornaments on the table in the shape of trees, demonstrating the modern flavour with the fashionable style.

空间的色彩采用素雅的色调突出现代简约的装饰风格,空间的装饰通过吊灯的设计和桌上树木造型的摆设来表现,时尚中更显现代气息。

202 SOFT DECORATION: FABRICS IN HOME DESIGN

The vestibule in the corridor uses the mirror-tinted glass as the partition wall, on which there is a stone portrait, solid and stately, radiating dignity and a feeling of reverence.

设置在房间走廊的玄关,用茶色玻璃镜面作为割断,在上面摆放的一尊石质人像,坚实而又厚重,给人以威严和敬畏之感。

Here the metal candlestick is a useful furnishing item. The candles light up the space and easily bring us into a nostalgic mood.

金属烛台作为陈设品,在照亮空间的同时也很容易将人们的思想带入怀旧的情绪当中。

The intercultural collision in the space between different cultures, represented by integrated elements such as the pottery jars with intense traditional flavour, the floor lamp with strong contemporary feelings and the seats weaved with iron wires, leads people to experience the abundant cultural aura.

这个空间将传统气息浓重的陶罐、现代感强烈的地灯以及由铁丝编制的座椅等元素混合在一起，形成不同文化间的碰撞，进而让人感受到强烈的文化气息。

Fabrics
织物

Fabrics in interior design include window curtains, beddings, cushions, table cloths and carpets. They are functional as well as decorative elements in home spaces, and play an important role in soft decoration.

Colour, pattern and texture are the three key elements in fabric selection. Fabrics visually satisfy our psychological and physiological needs with colours. Fabrics with different patterns create different decorative effects. Textures are also important for fabrics. Rough or smooth, brushy or lumpy, machine-weaved or handmade, various textures could help enhance decorative effects of fabrics in home design.

室内织物指窗帘、床罩、靠垫、台布、地毯等既有实用价值又能起装饰效果的家居纺织物。装饰织物在室内装修中起着非常重要的作用，是家居软装饰设计中重要的装饰细节。

织物的颜色、图案、布料的选择直接影响空间的装饰效果。织物色彩设计需要注意通过人们对色彩的视觉感受来满足人们的心理和生理需求。不同图案的织物对居室产生不同的效果。材质肌理也是影响织物的重要因素，粗糙与细腻、毛茸与疙瘩肌理的协调美，通过机织、手工制作，达到尽善尽美的地步，将织物的装饰效果体现得淋漓尽致。

The soft decoration in this bedroom is a blend of lighting, textile and green planting. The elegant space feels cosy and relaxing.

空间的软装饰融合了灯光、织物、绿植等元素,淡雅中更显舒适、放松的气氛。

Here is an example of the combination of pattern and colour. In the bedroom, the images of leaf on the pillow and on the screens and the images of orchid on the cupboards form a green theme. Such details complete a harmonious bedroom.

将图案和色彩设计相结合突出树叶的主题，卧室空间中将枕头的织物图案与屏风上的图案以及床头柜的兰花结合在一起，构成一个绿色的主题，成为卧室空间中的装饰细节。

The cushions with different colours, as well as the bright flowers in the vase on the coffee table, exemplify the use of colour in soft decoration.

靠垫采用不同的色彩作为装饰元素,与茶几上的鲜艳的插花相呼应,将空间的软装味道渲染得浓郁而又强烈。

Fabrics play a key role in soft decoration and the fabric design in the bedroom in particular. The space of the bedroom is the place where people can sleep and have a rest; therefore it needs to be designed with a warm and gentle feeling. The increased proportion of fabrics in the bedroom space can make people totally relaxed both physically and mentally, leaving them a safe and comfortable feeling. There are lots of choices of fabrics, namely, cotton, hemp and silk, which can all be applied to the soft decorative design of the bedroom.

在软装饰设计中织物始终占据重要的角色，尤其是卧室空间的织物设计。卧室空间是人们睡觉、休息的场所，因此需要温暖和柔软的感觉，在卧室空间增添织物装饰的比例能够给人以安全、舒适的感觉，使人能够得到全身心的放松。织物的材质选择，可以有多种，棉、麻、丝等都可以应用到卧室的软装设计当中。

The black cushion with floral patterns set off by the valance and the bed sheet in romantic white seems so smart. The white valance at the bedside against the background of the contrast between white and black endows the space with nebulosity and gentleness.

在浪漫的白色帷帐和白色床面上,黑色花面的靠垫显得很俏皮。在色彩上黑白又形成了对比色,白色的床头帷幔将空间渲染得朦胧而又柔软。

Texture
肌理

Texture in soft decoration refers to that of natural materials such as timber and stone, with natural patterns which act as decorative elements in home spaces. Such textures are important details in home design. With the rising awareness of saving natural resources, artificial materials such as man-made boards and marble are extensively used as important textures in soft decoration.
Material surfaces with different textures create different tactile and visual experiences. Texture, in terms of visual art, emotionally enriches a space just as colour does.

家居软装饰中的肌理指的是自然材料,如木、石等自然形成的现实纹理以及由此产生的空间装饰效果。肌理同样是家居软装饰设计中的装饰细节。随着时代的发展及自然材料的逐渐稀缺,通过人工处理的肌理如人造板材、仿大理石等同样也是软装饰中的肌理。
材料表面的纹理、组织构造能够给人以触觉质感和视觉触感,作为视觉艺术的一种基本形式,肌理同色彩一样具有表达情感的功能,能够增加空间富于多变的特征。

SOFT DECORATION: FABRICS IN HOME DESIGN

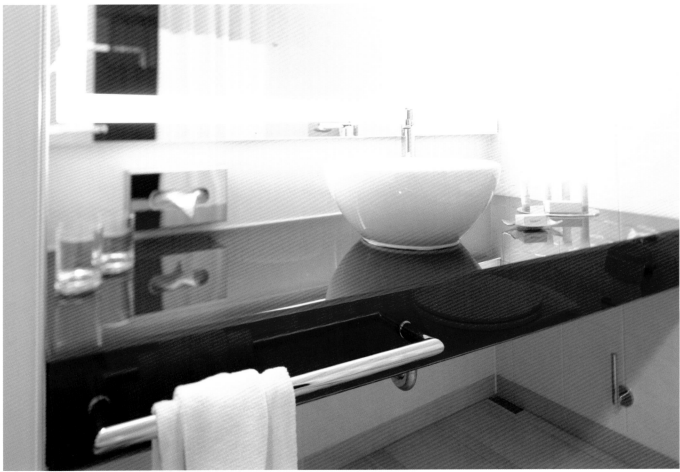

The lighting design fills the room with the visual contrast between the cold and the warm tones.

用灯光设计使空间形成冷、暖对比的视觉感受。

The marble and the white ceramic tiles create a white space which is pure and bright.

用大理石和白瓷砖将空间营造出一个白色的世界,纯净、明亮。

Project 1 案例 1
Notting Hill, London
伦敦诺丁山别墅

Location: London, England
Designer: BROOSK SAIB
Photographer: Marcus Peel
Completion date: 2008

项目地点：英国，伦敦
设计师：布鲁斯克·赛博
摄影师：马库斯·皮尔
竣工日期：2008年

Project description:

This project in the trendy Notting Hill area of London beautifully combined contemporary aesthetics with stylish living. The idea behind this project was to create a bright, modern and comfortable home for a young couple about to start a family. These clients had a keen interest in Pop Art so they wanted the interior of their home to reflect this. Broosk used loud vibrant colours and interesting individual pieces of furniture in a matched "miss-match" style to compliment their art collection.

Broosk particularly enjoyed creating the bespoke kitchen for this project especially the beautiful Makassar ebony veneers for the exaggeratedly tall units, natural stone for the flooring and work surface to give it that slick but luxurious look. Added with very comfortable furniture and a large plasma TV, it is clear to see why this is the clients' favourite room in the house.

此项目位于繁华时尚的伦敦诺丁山地区，设计将现代美与潮流生活融为一体。项目的设计宗旨是为一对即将成立家庭的年轻夫妇打造一个明亮、现代而又舒适的家居环境。业主钟爱流行艺术，也希望房屋的室内设计能够体现流行艺术之美。布鲁斯克采用极富活力的色彩与有趣而又独特的家居陈设以混搭的方式展现了业主的艺术收藏。

在所有设计中，布鲁斯克尤其热衷于定制厨房的设计，特别是高高的组合橱柜上漂亮的孟家锡黑檀木面板、天然的石质地板及表面光滑而又极为奢华的工作台面。再加上温馨舒适的家具与宽大的等离子电视，就不难看出为什么在整栋房屋中业主对这个房间宠爱有加。

226 *SOFT DECORATION: FABRICS IN HOME DESIGN*

Project 2 案例 2

House
房子

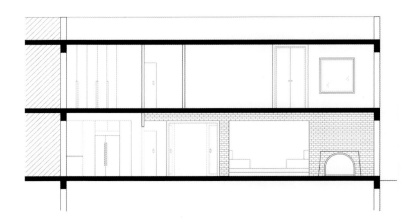

Location: Athens, Greece
Designer: Roxana Zouber
Photographer: Costas Picadas
Completion date: 2009
Area: 700 square metres
Built-up Area: 260 square metres

项目地点：希腊，雅典
设计师：罗克珊娜·佐伯
摄影师：科斯塔斯·皮卡德斯
竣工日期：2009年
场地面积：700平面米
建筑面积：260平方米

Project description:

The ethnic culture and nomadic spirit is apparent. Venetian mirrors, chandeliers made of Murano crystals unite with oriental couches create a very interesting result with the personal trademark of the owner. Roxana has formed a home that reflects her client's psychology resulting as the best exhibition of her work. Believing that harmony is vital in decorating interior space, attention has been given to scale, materials, style and details for the result to impress and to calm. The house with its high ceiling, large balcony windows that open onto a pristine garden, obtains warmth due to colourful fabrics, oriental cushions, carved mirrors and doors that at times function as tables or floor screens and others as entry into "mystical passages". The space is dominated by grandiose materials like crystal, wood and mosaics.
Primarily the house is designed in glass, concrete and wood. Mosaics and ceramic handmade tiles are used in the patio and pond areas.
The ground floor houses the living room, the elevated dining area and kitchen, which look out onto the garden and pond area. The pond is surrounded by an almost theatrical environment with ethnic brushstrokes, oriental and Victorian vases, and other colourful objects, as a continuation of the interior that is formed in the same style.

此设计中民族文化与游牧精神显而易见。威尼斯镜子、穆拉诺岛水晶制成的枝形吊灯与东方风格的长沙发搭配组合，营造了一个极具主人个人特色的十分有趣的氛围。罗克珊娜此项设计中最大的亮点就是房屋设计反映了业主的心理特征。设计师坚信和谐是室内装饰中极为重要的因素，因此在房屋比例、装饰材料及细节上下了很大工夫，以给人留有深刻印象并保持厚重沉稳的设计风格。高高的天花板、宽大的阳台窗与复古式的花园形成开放的布局。房子通过色彩鲜艳的织物、东方风格的靠垫、饰有雕刻花纹的镜子与门来获取热量，门有时被用作桌子或屏风，有时充当通往神奇通道的入口。房屋设计主要采用水晶、木头及马赛克等富丽堂皇的装饰材料。
房屋的主要部分的建筑材料为玻璃、混凝土与木头。露台与池塘则采用了马赛克与手工陶瓷瓷砖等原材料。
一层空间布局主要有客厅、突出的餐厅空间与厨房，厨房与花园和池塘相互开放。池塘周围环境极富戏剧性，比如，具有民族特色的绘画、凝聚东方特色与维多利亚风格的各式花瓶及其他色彩斑斓的装饰物，以此来作为室内空间的延续，与室内风格相互呼应。

230　SOFT DECORATION: FABRICS IN HOME DESIGN

Project 3 案例 3

The Manor, 10 Davies Street

豪华庄园大街10号住宅公寓

Location: London, UK
Designer: Casa Forma
Completion date: 2010
Area: 260 square metres

项目地点：英国，伦敦
设计师：Casa Forma室内设计工作室
竣工日期：2010年
室内面积：260平方米

Project description:

Casa Forma adds another bespoke luxury property to its collection of high-end residential properties in London. Set across 260 square metres, the splendid lateral four-bedroom flat spreads across the 3rd floor of a Georgian pottered building offering a discerning choice of residence appropriately called "The Manor". Perfectly placed in the heart of the prestigious and elegant streets, Davies Street is discreetly located between the Grosvenor Square and Bond Street in Mayfair, London's most exclusive district neighbouring well-known restaurants and boutiques.

Casa Forma has designed the architecture through an extensive amount of stripping-out and demolition work in all rooms to create unique lateral space with uninterrupted views.

The floor has been designed with exceptional materials to create a sophisticated environment using an outstanding selection of natural materials including a mixture of stained walnut, ebonised and ebony macassar timbers. Moon Onyx clad to the entrance foyer walls with back lighting bestows a subtle glow, and the bronze trims designed to frame the onyx seem as if it were a piece of art. The setting completely reflects understated luxury, very true to Casa Forma's design philosophy.

The entrance floor leading to the corridor consists of a natural rare off-white stone with trims and details in bronze. These minutiae have also been used in other areas to bring about the same concept. Melted bronze and silver sheets of metal clad to feature walls work in complete harmony with the ebonised macassar panelling. Other walls are finished in polished plaster and upholstered with calf skin. The approach to the bedrooms has also been kept subtle in tones of colour to create a textured feature wall behind the beds. Traces of walnut and ebonised macassar have been used to design the bespoke joinery. Radiator cases have been cleverly concealed behind panelling in all rooms.

The en-suite bathrooms use the same tones of colour with rare natural stone and antique gold bathroom sanitary ware. The unrivalled combination of distinct architecture,

Casa Forma设计的这个工程一直有大量的装修施工。为了最大程度打造宽广的视野空间，他们在每一个房间都进行了大面积的拆除工作，以期在各个房间之间打造出连贯的空间。

Casa Forma工作室采用精湛的建筑工艺及丰富多样的自然材料对室内地板进行了精心的设计，使用了胡桃木、乌木和黑檀木等木材。背光处理的月光缟玛瑙石板让入口门厅的墙壁散发着柔和的光彩，青铜镶边精心勾勒出缟玛瑙石的造型，使其看起来像一件艺术品。这种完美的设计风格充分地展现出了Casa Forma工作室一贯低调奢华的设计理念。

走进入口门厅可以发现一条通往走廊的罕有白色天然石板，上面有青铜镶嵌的精致装饰细节。这些细节设计同样出现在了其他房间里，以呈现统一的设计主题。熔融青铜和银色金属薄片装饰的特色墙面与乌木贴面镶板完美相融。其他墙面采用了抛光石膏饰面，并加以小牛皮加以装饰。卧室的设计手法亦十分精巧，在床头后面打造了一堵具有独特纹理的特色墙面。所有房间的细部装饰巧妙地隐藏在镶板后面。

套间内的浴室采用同样的色调，配有罕见的天然石材和古色古香的金色卫浴设备。独特的设计风格，精良的品质，精湛的工艺以及对细节的用心，这些成就了Casa Forma工作室的此次设计项目。凭借大量精致的手工和精湛的设计，Casa Forma工作室把这套公寓改造而出，成为位于伦敦上层社区梅菲尔豪华环境中的一间非凡住宅，与之所在的上流街区的氛围达成了新的和谐。

238 *SOFT DECORATION: FABRICS IN HOME DESIGN*

Project 4 案例 4

Villa No.10
10号别墅

Location: Shanghai, China
Designer: MoHen Design International /Hank M. Chao
Participant: Ruby Zhao
Photographer: MoHen Design International / Maoder Chou
Completion date: 2010
Area: 430 square metres

项目地点：中国，上海
设计师：牧桓国际室内设计事务所
（主设计师：赵牧桓，赵路参与设计）
摄影师：牧桓国际室内设计事务所（周宇贤）
竣工日期：2010年
项目面积：430平方米

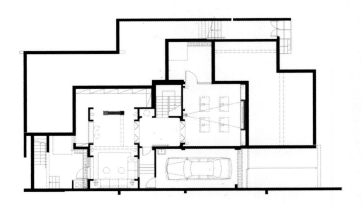

Project description:

At the ground floor entrance, the designers made a cut of the opening space on the vertical elevation which is directly connected to the first floor corridor. It makes the very first entrance hall turn into a place which seems more interesting. Such a vertical integration can also allow a glance from the first floor corridor to the ground floor. From the functional point of view, it is more practical as well. Around each side is the living room and dining kitchen. Facing to the entrance, there is a high grey space leading to the top floor balcony. Here the designers adopted a concept of ancient Chinese architecture, but they just put it directly to the entrance porch. It is really rare to have such a space in a modern villa, and the space thus got its own style. There is no need for more complex design on garden landscaping, because lighting itself is very exciting already. The designers put some bamboo inside to establish a certain atmosphere. As to the living room, they brought in the concept of "a depositary of Buddhist texts": apart from the high ceiling, they made some bookshelves on the mezzanine, and also built some bridges as a way to connect spaces. The living room thus looks somewhat like a study.

After all, the original building framework of the first floor limits certain parts. Only for the layout of the bathroom, the designers adopted more flexible space combination to make this part more bright or private. Some structures which originally cannot be made as a shower space now become opened instead. Now at the master bedroom on the second floor, the designers deliberately cut corners in the space between the balcony and the bedroom to form the whole glass connection to make the external sight wider. For the background of the master bed, they used shell plate positioning to emphasize its own specificity. They decorated the entrance of the changing room with a traditional Chinese painting to make a dull door into an opaque paint. The other side of the master bedroom is master's internal study room which directly connects to the master bathroom. To define the shower space and toilet room, they used the ways of circle geometry on the layout of bathroom. The function of the basement is mainly to meet the master's needs for entertainment; a large living room combines functions of wine collection and tasting, as well as visitor receiving. The space of bar and wine cellar are designed to be open so that they won't appear to be too narrow.

在一层入口处，设计师在垂直的立面上做了一个空间切割的开口直接连接二楼的廊道。这样，第一进位的入口大厅显得更加引人注目。这样的垂直整合设计手法可以使人们从二楼走廊就可以看到一楼人流进出的情况。从功能性角度考虑，这样的设计更具实用性。左右两旁是客厅和餐厅厨房。入口对面的灰色空间直通顶楼阳台，此处设计师借用了中国古代建筑的设计手法，因为这么垂直狭长而又有自然光线的进光的取景实在太难得出现在现代别墅的设计里了，放进来反而显得有些味道。通过这种手法，花园景观已无需更多的复杂设计，因为灯光本身就是一道亮丽的风景。设计师在室内安放了一些竹子，形成了一种浓郁的意境。在客厅的设计中，除了天花板的特殊设计之外，设计师还引入了佛鱼藏经阁的概念，不仅在挑高上做了处理，夹层部分做了书架并搭接了桥梁作为连接空间的对接方式，客厅倒也引入几分书卷气息。

二楼毕竟原本的建筑框架已经具有一定的局限性，只有在卫生间的格局上采取了一些比较灵活和闭合的空间结合方式让局部更加敞亮或者私密，有些原本无法做淋浴间的格局反而开放了。到了三楼的主卧，在卧室部分跟阳台的衔接设计师特意把转角切口全让出来做玻璃连接，让对外视线能够更宽一点，主床的背景则用了贝壳板定位强调了一下特殊性，更衣室的入口在装饰面上则用了鲤鱼国画的方式处理，让原本无趣的门片成为一幅可以透光的画。主卧的另外一头则是主人的内部书房，书房直接连接卫生间，卫生间在格局上倒是应用了一点几何圆圈的方式来界定淋浴和马桶间。地下室主要尽量满足主人的娱乐需求，一个大的起居室结合了其藏酒和品酒以及会客的功能，在吧台和藏酒窖的部分则尽量适度开放，让空间的层次感能够较为明显而且不会显得狭小。

Index 索引

1. Project: Blue Sky Home
   Designer: o2 Architecture
   Completion: 2009
   Location: Yucca Valley, USA
   Photographer: Nuvue Interactive
   Area: 1,000 SF

2. Project: Thong Lor Villa
   Designer: Monochrome Inc. Interior Design
   Managing Director: Andrew Loader
   Completion: 2009
   Location: Thong Lor, Bangkok, Thailand
   Photographer: Tree Studio
   Area: 1,000 SF

3. Project: 45 Faber Park
   Designer: Ong&Ong Pte Ltd
   Completion: 2009
   Location: Singapore City, Singapore
   Photographer: Derek Swalwell, Tim Nolan
   Area: 592 m^2

4. Project: Family Residence in Vilnius
   Designer: Indra Marcinkeviciene
   Completion: 2006
   Location: Kaunas, Lithuania
   Photographer: Valdas Račyla
   Area: 25,428 m^2

5. Project: Apartment in Kaunas
   Designer: Indra Marcinkeviciene
   Completion: 2008
   Location: Kaunas, Lithuania
   Photographer: Valdas Račyla
   Area: 9,238 m^2

6. Project Name: Dahua Aqua Island
   Designer: Fang Huang Design Studio
   Completion: 2008
   Location: Chengdu, China
   Photographer: Wang Jianlin
   Area: 178.9 m^2
   Main Material: floor board, stone, mosaic, silver mirror, emulsion paint

7. Project: The Orange Grove
   Designer: Indra Marcinkeviciene
   Completion: 2009
   Location: Orchard Road, Singapore
   Photographer: Albert Lim (Albert Lim Photography) and Rory Daniels (Rory Daniels Photography)

8. Project: Casa Mataró
   Designer: Elia Felices interiorismo
   Completion: 2008
   Location: Mataro, Spain
   Photographer: Rafael Vargas
   Area: 190 m^2

9. Project: Alma Road Residence
   Designer: Kerry Phelan
   Completion: 2008
   Location: Alma Road, St Kilda
   Photographer: Shannon McGrath Photography

10. Project: Casa K+JC2
    Designer: SPACE / Juan Carlos Baumgartner
    Completion: 2008
    Location: Jardines del Pedregal, México
    Photographer: Pim Schalkwijk
    Area: 600 m^2

11. Project: The Cortazzo Ranch
    Designer: Martyn Lawrence Bullard
    Completion: 2009
    Location: Malibu, California, USA
    Photographer: Tim Street Porter
    Area: 600 m^2

12. Project: Edge House
    Designer: Martyn Lawrence Bullard
    Completion: 2008
    Location: Boulder, Colorado, USA
    Photographer: Tim Street Porter
    Area: 600 m^2

13. Project: Maison Moschino
    Designer: Moschino
    Completion: 2010
    Location: Milan, Italy
    Photographer: Massimo Listri
    Martina Barberini
    Henri Del Olmo
    Area: 3,000 m²

14. Project: Apartment Formica
    Designer: Héctor Ruiz-Velázquez MARCH
    Completion: 2010
    Location: Madrid, Spain
    Photographer: Pedro Martinez
    Area: 70 m²

15. Project: The Samling
    Designer: Andrew Onraet
    Completion: 2010
    Location: Cumbria, England
    Photographer: von Essen hotels

16. Project: Penthouse Downtown Montreal
    Designer: RENÉ DESJARDINS
    Completion: 2010
    Location: Montreal, Canada
    Photographer: André Doyon
    Area: 3,300 SF

17. Project: ENCARNACION
    Designer: RENÉ DESJARDINS
    Completion: 2007
    Location: Madrid, Spain
    Photographer: Antonio Terron
    Area: 160 SF

18. Project: Residential Villa
    Designer: Michael Clattenburg
    Completion: 2009
    Location: Dubai, United Arab Emirates
    Photographer: Andrew Garner
    Area: 1,700 m²

19. Project: Yapu Sample Apartment
    Designer: Mark Lintott, Vivian Lee, Mei Yi Chou
    Completion: 2008
    Location: Da Zhi, Taipei
    Photographer: Marc Gerritsen
    Area: 230 m²

20. Project: Shangri-La
    Designer: B+H CHIL Design
    Completion: 2010
    Location: Vancouver, Canada
    Photographer: Ed White

21. Project: Granite Lodge at the Ranch at Rock Creek
    Designer: Jet Zarkada, Los Griegos Studio, LLC
    Completion: 2010
    Location: Montana, USA
    Photographer: Los Griegos Studio, LLC, the Ranch at Rock Creek
    Area: 53,800 SF

22. Project: The Waterfall House
    Designer: Andres Remy, Flavia Bellani, Marcos Pozzo, Paula Mancini, Lauru Rodriguez Segat, Rodri guez Leandro Llebana
    Completion: 2006
    Location: Buenos Aires, Argentina
    Photographer: Andres Remy Architects
    Area: 340 m²

23. Project: Urban Farmer
    Designer: d-ash design
    Completion: 2008
    Location: Portland, USA
    Photographer: Michael Mathers
    Area: 3,690 SF

图书在版编目（CIP）数据

宅妆：家居软装饰 / 杜丙旭编；代伟楠，李婵译. -- 沈阳：辽宁科学技术出版社，2012.10
ISBN 978-7-5381-7489-2

Ⅰ. ①宅… Ⅱ. ①杜… ②代… ③李… Ⅲ. ①住宅—室内装饰设计—图集 Ⅳ. ①TU241-64

中国版本图书馆CIP数据核字(2012)第099147号

出版发行：辽宁科学技术出版社
　　　　　（地址：沈阳市和平区十一纬路29号　邮编：110003）
印　刷　者：利丰雅高印刷（深圳）有限公司
经　销　者：各地新华书店
幅面尺寸：240mm×280mm
印　　张：16
插　　页：4
字　　数：50千字
印　　数：1～2000
出版时间：2012年 10 月第 1 版
印刷时间：2012年 10 月第 1 次印刷
责任编辑：陈慈良
封面设计：曹　琳
版式设计：曹　琳
责任校对：周　文
书　　号：ISBN 978-7-5381-7489-2
定　　价：228.00元

联系电话：024-23284360
邮购热线：024-23284502
E-mail: lnkjc@126.com
http://www.lnkj.com.cn
本书网址：www.lnkj.cn/uri.sh/7489